The Legacy of Albert Einstein

A Collection of Essays in Celebration of the Year of Physics

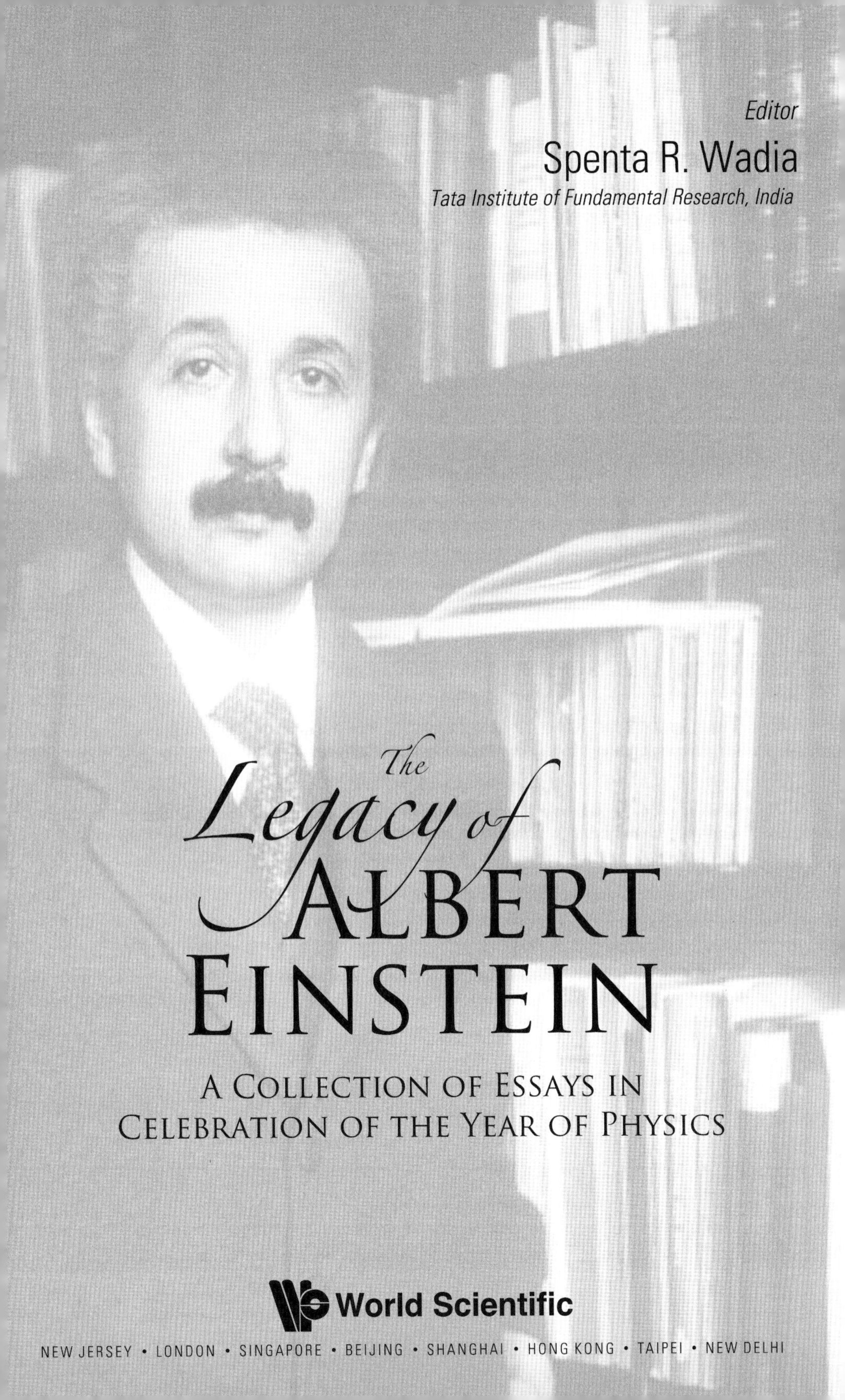

Editor

Spenta R. Wadia

Tata Institute of Fundamental Research, India

The Legacy of ALBERT EINSTEIN

A COLLECTION OF ESSAYS IN CELEBRATION OF THE YEAR OF PHYSICS

World Scientific

NEW JERSEY • LONDON • SINGAPORE • BEIJING • SHANGHAI • HONG KONG • TAIPEI • NEW DELHI

Published by

World Scientific Publishing Co. Pte. Ltd.
5 Toh Tuck Link, Singapore 596224
USA office: 27 Warren Street, Suite 401-402, Hackensack, NJ 07601
UK office: 57 Shelton Street, Covent Garden, London WC2H 9HE

British Library Cataloguing-in-Publication Data
A catalogue record for this book is available from the British Library.

Cover Credit: Courtesy of Leo Baeck Institute, New York.

THE LEGACY OF ALBERT EINSTEIN
A Collection of Essays in Celebration of the Year of Physics

ISBN 981-270-049-8
ISBN 981-270-480-9 (pbk)

Printed by FuIsland Offset Printing (S) Pte Ltd, Singapore

****

Dedicated to the memory of Homi Jehangir Bhabha (1909–1966) —
Scientist, Artist and a Visionary builder of scientific institutions in India

****

Head of Einstein, Jacob Epstein, 1955 (Courtesy: Tata Institute of Fundamental Research, Mumbai)

PREFACE

The Legacy of Albert Einstein (1879–1955)

****

The epoch making 1905 papers of Albert Einstein, mark a turning point in the history of physics and also the history of mankind. In this respect he shares a platform with Galileo and Newton who gave us the basic formulation of mechanics in terms of forces and accelerations, and with Faraday and Maxwell, who introduced the notion of fields in space and time.

In this hundredth anniversary year we look back and recall the main threads Einstein wove into the tapestry of physics. This collection of articles, *The Legacy of Albert Einstein*, presents developments whose initiation can be found in the works of Einstein. We have not restricted ourselves to the 1905 papers, but include all his major contributions to physics. Since Einstein was a major public figure of the 20th century we also include this aspect in the collection.

One cannot fail to note that one of the most important legacies of Einstein, is the fact that he did not make any conscious value differentiation between different areas of physics. He was equally at home with relativity, geometry, radiation theory, Brownian motion, statistics, molecular physics and so on. There was his quest for frameworks and general principles within which the laws of nature operate, and there was an equally important quest to see how these laws manifest themselves in the world around us. In a deep sense these activities are inevitably mixed up in the grand enterprise to understand our world.

We present a glimpse of the scientific contribution of Einstein, which will hopefully provide an overall background to the articles in this book.

1. Einstein's Annus Mirabilis: Five Papers that Changed the World

The 1905 papers sowed the seeds of both revolutions of 20th century physics: Relativity and Quantum Theory. Before Einstein there were Newton's laws of mechanics and Maxwell's equations of electrodynamics. Thermodynamics was a well established subject and there was Boltzmann's microscopic formula for entropy.

- Brownian motion and the reality of atoms:
  Einstein's doctoral dissertation on molecular dimensions, followed by his famous statistical formula for the motion of suspended 'Brownian' particles, gave a beautiful method of calculating Avogadro's number and the size of molecules. Using this framework, the existence of the underlying molecular structure of matter was experimentally concluded beyond doubt, notably by the work of Jean Perrin. Satya Majumdar summarizes Einstein's original work and its applications to physics, probability theory and computer science.
- The light quantum hypothesis:
  Einstein put forth the daring hypothesis, that light has particulate properties ('a kind of molecular structure in energy'). His heuristic principle stated that light is created and annihilated in discrete quanta in its interaction with matter. He suggested testing his proposal using the photoelectric effect. This was a revolutionary idea, because it was in absolute contradiction with Maxwell's theory of the continuously varying electromagnetic field that describes pure radiation. His conviction in the verity of thermodynamics and in Boltzmann's formula led the way. There is a possibility that his work on the discrete nature of matter influenced his work on the discrete nature of the electromagnetic field. Virendra Singh reviews the history of the light quantum.
- Special relativity:
  'The electrodynamics of moving bodies', is the paper on special relativity. In one fell swoop, Einstein replaced all mechanical explanations (and other attempts) of the constancy of the speed of light, irrespective of the motion of the source, by asserting a symmetry principle which all laws of nature must conform to! Thus was born special relativity and a new kinematic framework for physics. Maxwell's equations are true but Newton's mechanics had to be replaced. At the age of 16, while in Arrau, Einstein had asked, 'If I pursue a beam of light with velocity c, I should observe such a beam of light as a spatially oscillatory electromagnetic field

at rest'.[1] This mystery was resolved by special relativity: one is never at rest with respect to light! Subsequently, he derived his celebrated $E = mc^2$, which is probably the most popular scientific formula of our times.

2. Einstein's Magnum Opus: General Relativity

The special relativity paper has no references! There is an apparent ease with which conclusions emerge in this paper. The logic seems flawless and unbreakable. Special relativity deals with special reference frames in constant uniform motion with respect to each other and also does not incorporate the instantaneous law of gravitation. Einstein set out to rectify these seemingly unrelated shortcomings.

In 1907, in his own words,[2] 'Now it came to me: The fact of the equality of inert and heavy mass, i.e. the fact of the independence of the gravitational acceleration of the nature of the falling substance, may be expressed as follows: In a gravitational field (of small spatial extension) objects behave as they do in a space free of gravitation, if one introduces in it, instead of an 'inertial system', a reference system which is accelerated relative to an inertial system. Einstein calls this the happiest thought of his life. In it, he identified the gravitational force with accelerations: the fall that we attribute to the earth's gravity is no different from the forward fall we experience when we jam the brakes of a car!

The next big step in 1912 was the realization that spacetime is curved and not flat! The search for the correct, generally covariant, equations of general relativity in the framework of Riemannian geometry took many more years. The final version of the field equations of gravitation were presented to the Prussian Academy on 25 November, 1915. They are $R_{ij} - \frac{1}{2}Rg_{ij} = T_{ij}$.

The discovery of these equations was an extraordinary struggle, fraught with errors and corrections. It is very encouraging and inspiring to see how the great Einstein struggled to discover these equations. He did not have the benefit of knowing the deep underlying symmetry principle hidden in the equations he was to discover! This realization was almost postfacto. The symmetry followed from the equations rather than the other way around, as

[1] Autobiographical notes, in *Albert Einstein: Philosopher Scientist* edited by P. A. Schilpp, Library of Living Philosophers, Vol. 7, 1949, p. 53.
[2] Autobiographical notes, p. 67.

most textbooks on general relativity have us believe. If his 1905 papers have a similarity with the style of Mozart, the development of general relativity reminds one of Beethoven. In a lecture to the University of Glasgow in 1933, Einstein said[3] 'The years of searching in the dark for a truth that one feels but cannot express, the intense desire and the alternations of confidence and misgiving until one breaks through to clarity and understanding are known only to him who has experienced them'.

As is well known, the predictions of the general theory of relativity, concerning the bending of light, the perihelion of mercury and the redshift, were eventually vindicated by experiment. Subsequently, Einstein went on to find that his equations support gravitational waves with two polarizations and discovered the quadrupole formula. In 1917 he proposed a cosmological solution by including the cosmological constant (dark energy in today's parlance). This work laid the foundations of cosmology in the framework of general relativity. Jayant Narlikar traces the development of cosmology. Subir Sarkar gives an observational perspective of modern cosmology, and points to the fundamental problems of dark matter and dark energy. B. Sathyaprakash reviews the current and future status of observing gravitational waves, which bear the imprints of the earliest universe and also other strongly gravitating objects.

Soon after, the field equations of general relativity were presented to the world, Schwarzschild found his famous blackhole solution. Atish Dabholkar traces the development of blackhole physics and the crucial role it plays in the discovery of the quantum structure of gravity.

The focusing properties of the gravitational field were discovered by Raychaudhuri in a fundamental contribution. The Raychaudhuri equation is a precursor to the singularity theorems of Penrose and Hawking. General relativity was also the main inspiration for Yang and Mills to discover non-Abelian gauge theories in 1957.[4]

3. Contributions to Condensed Matter, Optics and Quantum Mechanics

It was characteristic of Einstein that he mused about all basic physics problems of his time. In 1906, he applied the quantum hypothesis, outside of

[3] *Subtle is the Lord: The Life and Science of Albert Einstein*, Abraham Pais, Oxford Univ. Press, 1982, p. 257.

[4] Chen Ning Yang, *Selected Papers 1945–1980 with Commentary*, W. H. Freeman and Co.

radiation theory, to calculate the specific heat of solids. This is the first paper on the quantum theory of the solid state. At the first Solvay conference in 1911, Einstein's concluding talk summarized, 'The Current Status of the Problem of Specific Heats'. T. V. Ramakrishnan, gives an account of the development of condensed matter physics and discusses the present status of some aspects of this vast area of physics.

In 1916, after general relativity, he once again turned to radiation theory and statistical fluctuations. He discussed spontaneous and induced radiative processes. The concept of the photon as a particle with a quantum of energy and momentum was established in the same year. There is a gap of eleven years before Einstein associated a momentum $p = \frac{h\nu}{c}$ with the light quantum! The concept of the photon remained controversial until irrefutable evidence was provided by Compton's experiments in 1923. The photon was systematically formulated in the quantum theory of the electromagnetic field by Dirac in 1930. Shortly after that, Heisenberg and Pauli began a systematic treatment of relativistic quantum field theory, and its successful application to quantum electrodynamics was done by Feynman, Schwinger and Tomonaga in the late 1940s.

In 1924 Bose's attempt to derive Planck's radiation law, using photons as indistinguishable particles, led to Bose–Einstein statistics for indistinguishable particles and the discovery of the phenomenon of Bose–Einstein condensation. Narendra Kumar writes about this contribution, and details modern experimental achievement of the new Bose–Einstein phase of matter and its impact on basic and condensed matter physics.

Even though Einstein was a pioneer of quantum theory and was the first to recognize that the new mechanics has a wave-particle duality, he was never convinced that the quantum mechanics formulated by Heisenberg, Schroedinger, Dirac and Born was a fundamental description of reality. The statistical interpretation of quantum mechanics greatly troubled him and he believed that a more fundamental theory underlying quantum mechanics had to be discovered. In 1935, with Podolsky and Rosen he came up with a thought experiment that reveals the conflict between locality and quantum entanglement. According to Einstein, 'No reasonable definition of reality can be permitted to do this'. However the present experimental verdict is not on Einstein's side. Virendra Singh discusses these developments.

4. Unified Field Theory

In the later years of his life, beginning around 1920, Einstein was mainly preoccupied with a quest for a unified theory of gravitation and electromagnetism. Even for the equations of general relativity, he had the following to say, '...it was essentially not anything more than a theory of the gravitational field, which was somewhat artificially isolated from a total field of yet unknown structure'.[5] He was appreciative of the work of Kaluza which achieved unification using a small fifth circular dimension, but (as usual) he pursued his own thoughts.

Einstein did not succeed in his unification program, but in the decades that followed this idea of a unified theory of all forces remained an inspiration and culminated in the unified theory of weak and electromagnetic interactions of Glashow, Weinberg and Salam in the 1970s. Unification of gravity with the weak, electromagnetic and strong forces is still one of the central themes in physics. David Gross, Michael Atiyah and Ashoke Sen trace these developments and describe the development of string theory as a framework for a unified theory. Abhay Ashtekar summarizes efforts to quantize gravity and also describes developments in loop quantum gravity.

5. Einstein's Persona:

During his lifetime, Einstein was a public figure. He reached this stature by the sheer revolutionary nature of his science and his historic achievement. At the age of 37 years, he was the greatest scientist in the world. He was an active champion of peace, individual freedom and social justice. Unlike any other scientist of his generation he took strong and principled political positions. Even though he had supported nuclear weapons as a defense against Naziism, after the second World War, he publicly opposed and campaigned for a world free of nuclear weapons. In a solemn speech in 1950, when the hydrogen bomb was being developed, he said, 'a weird aspect of this development lies in its apparently inexorable character. Each step appears as inevitable consequence of the one that went before. And at the end, looming ever clearer, lies annihilation'. Fifty-five years later the world is still unsafe! T. Jayaraman, in his article, delineates Einstein's activism and involvement with social and political issues of his times.

---

[5] Autobiographical notes, p. 75.

Finally, in the words of Abraham Pais,[6] 'Einstein was the freest man I have known. He was a master of his own destiny. His deep sense of destiny led him farther than anyone before him. It was his faith in himself that made him persevere'.

Acknowledgments

I would like to thank P. Balaram, Editor of *Current Science*, for inviting me to guest edit the special section of Current Science (25 December, 2005) on the *Legacy of Albert Einstein*. This book is mainly based on it. My thanks, to all the authors for their excellent contributions, to Leena Chandran-Wadia for discussions and for listening, to Miriam Intrator of the Leo Baeck Institute, New York, for correspondence, and to Lakshmi Narayan of World Scientific for her advice and help. The Einstein photos are courtesy of the Leo Baeck Institute, the Institute for Advanced Study at Princeton and the Tata Institute of Fundamental Research, Mumbai.

Spenta R. Wadia
Tata Institute of Fundamental Research
Mumbai, India
December, 2005

[6] *Subtle is the Lord*, p. 17.

CONTENTS

****

Preface ix
Spenta R. Wadia

Einstein and the Search for Unification 1
David Gross

Einstein and Geometry 15
Michael Atiyah

String Theory and Einstein's Dream 25
Ashoke Sen

Black Hole Entropy in String Theory: A Window into the Quantum Structure of Gravity 47
Atish Dabholkar

The Winding Road to Quantum Gravity 69
Abhay Ashtekar

Brownian Functionals in Physics and Computer Science 93
Satya N. Majumdar

Bose–Einstein Condensation: Where Many Become One and So There is Plenty of Room at the Bottom 131
N. Kumar

Many Electrons Strongly Avoiding Each Other: Strange Goings On 149
T. V. Ramakrishnan

Einstein and the Quantum 165
Virendra Singh

Einstein's Legacy: Relativistic Cosmology 193
Jayant V. Narlikar

Einstein's Universe: The Challenge of Dark Energy 207
Subir Sarkar

Gravitational Radiation — In Celebration of
Einstein's *Annus Mirabilis* 225
B. S. Sathyaprakash

Albert Einstein: Radical Pacifist and Democrat 251
T. Jayaraman

CHAPTER 1

Einstein and the Search for Unification

****

DAVID GROSS

Department of Physics, University of California,
Santa Barbara, CA 93106, USA

Einstein spent the last thirty years of his life searching for a unified field theory. I discuss Einstein's attempts at unification. I examine his mistakes, ask why he went wrong, and wonder what might have happened if he had followed a slightly different route. I then discuss, very briefly, where we stand today in realizing Einstein's goals.

My topic is at the heart of Einstein's scientific life, the search for a unified theory of nature. This was Einstein's main pursuit for more than half of his scientific career. Most contemporaries viewed his attempts as a waste of time, a total failure or, at best, premature. But today we look with some admiration at his foresight. Having understood by the middle 1970's, to a large extent, all the four forces of nature in the remarkable successful standard model, attention has returned to Einstein's dream of unifying all the forces with gravity. The goal of unification has been at the forefront of fundamental physics for the last three decades.

In this article I shall, fully aware of the ease of hindsight, discuss Einstein's goals, his attempts to unify general relativity and electromagnetism, and to include matter. I shall discuss his mistakes, ask why he went wrong, and wonder what might have happened if he had followed a slightly different route. As I am not a professional historian I can get away with murder. I shall then discuss, very briefly, where we stand today in realizing Einstein's goals.

For many physicists, certainly me, Einstein is both a hero and a model. He stated the goals of fundamental physics, that small part of physics that probes the frontiers of physics in a search for the underlying laws and principles of nature. Einstein was a superb epigramist, who could capture in a single sentence many deep thoughts.

Here is his definition of the goal of the physicist:

> *The supreme test of the physicist is to arrive at those universal laws of nature from which the cosmos can be built up by pure deduction.*

I love this sentence. In one sentence Einstein asserts the strong reductionist view of nature: There exist universal, mathematical laws which can be deduced and from which all the workings of the cosmos can (in principle) be deduced, starting from the elementary laws and building up.

Einstein, more than any other physicist, untroubled by either quantum uncertainty or classical complexity, believed in the possibility of a complete, perhaps final, theory of everything. He also believed that the fundamental laws and principles that would embody such a theory would be simple, powerful and beautiful. The 'old one', that Einstein often referred to, has exquisite taste.

This exciting goal, which I first learned of when I was thirteen by reading popular science books, seemed to me so exciting that I vowed to become a theoretical physicist. Although I certainly had no idea what that meant, I did know that I wanted to spend my life tackling the most fundamental questions of physics. This goal led me to elementary particle physics in the 1960's and to string theory in the 1980's. This goal motivated Einstein to spend the last thirty years of his life in a futile search for a unified theory of physics.

Physicists are an ambitious lot, but Einstein was the most ambitious of all. His demands of a fundamental theory were extremely strong. If a theory contained any arbitrary features or undetermined parameters then it was deficient, and the deficiency pointed the way to a deeper and more profound and more predictive theory. There should be no free parameters — no arbitrariness.

Nature, he stated with confidence, *is constituted so that it is possible to lay down such strong determined laws that within these laws only rationally, completely determined constants occur, not constants therefore that could be changed without completely destroying the theory.* This is a lofty goal, under threat nowadays from those who propose the Anthropic principle, whereby many of the fundamental constants of nature, even some of the laws, are environmental in nature and might be different in different parts of the universe. For me and for many others however, this remains the ultimate goal of physics, and a guiding principle. A theory that contains arbitrary parameters, or worst of all arbitrarily finely tuned parameters, is deficient.

After his enormous success at reconciling gravity with relativity, Einstein was troubled by the remaining arbitrariness of the theoretical scheme. First, the separate existence of gravitation and electromagnetism was unacceptable. According to his philosophy, electromagnetism must be unified with general relativity, so that one could not simply imagine that it did not exist. Furthermore, the existence of matter, the mass and the charge of the electron and the proton (the only elementary particles recognized back in the 1920s), were arbitrary features. One of the main goals of a unified theory should be to explain the existence and calculate the properties of matter.

Before passing to a discussion of Einstein's attempts at unification I wish to make a remark concerning his work on special relativity in 1905, whose centenary we celebrate this year. One of the most important aspects of this work was to revolutionize how we view symmetry. Principles of symmetry have dominated fundamental physics in the 20th century, starting with Einstein in 1905.

Until the twentieth century principles of symmetry played little conscious role in theoretical physics. The Greeks and others were fascinated by the symmetries of physical objects and believed that these would be mirrored in the structure of nature. Kepler attempted to impose his notions of symmetry on the motion of the planets. The laws of mechanics embodied symmetry principles, notably the principle of equivalence of inertial frames, or Galilean invariance.

The symmetries implied conservation laws. Although these conservation laws, especially those of momentum and energy, were regarded to be of fundamental importance, they were regarded as consequences of the dynamical laws of nature rather than as consequences of the symmetries that underlay these laws. Maxwell's equations, formulated in 1865, embodied both Lorentz invariance and gauge invariance. But these symmetries of electrodynamics were not fully appreciated for over forty years or more.

This situation changed dramatically in the twentieth century beginning with Einstein. Einstein's great advance in 1905 was to put symmetry first, to regard the symmetry principle as the primary feature of nature that constrains the allowable dynamical laws. Thus the transformation properties of the electromagnetic field were not to be derived from Maxwell's equations, as Lorentz did, but rather were consequences of relativistic invariance, and indeed largely dictate the form of Maxwell's equations. This is a profound change of attitude. Lorentz, who had derived the relativistic transformation laws from Maxwell's equations, must have felt that Einstein cheated.

Einstein recognized the symmetry implicit in Maxwell's equations and elevated it to symmetry of space-time itself. This was the first instance of the *geometrization* of symmetry, and the beginning of the realization that symmetry is a primary feature of nature that constrains the allowed dynamical laws.

The traditional symmetries discovered in nature were global symmetries, transformations of a physical system in a way that is the same everywhere in space. Global symmetries are regularities of the laws of motion but are formulated in terms of physical events; the application of the symmetry transformation yields a different physical situation, but all observations are invariant under the transformation. Thus global rotations rotate the laboratory, including the observer and the physical apparatus, and all observations remain unchanged.

Gauge or local symmetry is of a totally different nature. Gauge symmetries are formulated only in terms of the laws of nature; the application of the symmetry transformation merely changes our description of the same physical situation, does not lead to a different physical situation. Today we realize that local symmetry principles are very powerful — they dictate the form of the laws of nature.

In 1912–17 this point of view scored a spectacular success with Einstein's construction of general relativity. The principle of equivalence, a principle of local symmetry — the invariance of the laws of nature under local changes of the space-time coordinates — dictated the dynamics of gravity, of space-time itself. Fifty years later gauge theories, invariant under local symmetry transformations, not of space-time but of an internal space of particle labels, assumed a central position in the fundamental theories of nature. They provide the basis for the extremely successful standard model, a theory of the fundamental, non-gravitational forces of nature — the electromagnetic, weak and strong interactions.

Surprisingly Einstein did not follow the symmetry route. He did not, in his attempts to unify physics, search for extensions of the symmetries that he had promulgated. If he had he might very well have discovered non-Abelian gauge theory or perhaps even supersymmetry. Why not follow this route that has dominated theoretical speculation in the latter half of the 20th century? I think the reason was that Einstein was unaware of the phenomenon of symmetry breaking. All of the new symmetries discovered in the latter half of the 20th century, that are at the heart of the standard model of particle physics and attempts at unification, are approximate, or are broken spontaneously, or hidden by confinement. It was only in the

1960s, and early 1970s, that these mechanisms of symmetry breaking were elucidated and the possibility of imagining new symmetries, not directly manifest in the world, but still dictating the dynamics, was possible.

For Einstein the existence, the mass, the charge of the electron and the proton, the only elementary particles recognized back in the 1920s, were arbitrary features. One of the main goals of a unified theory should be to explain the existence and calculate the properties of matter. When he contemplated his equation he distinguished between the left-hand side of the equation, which was a beautiful consequence of the profound symmetry of general coordinate transformations, and captures the curvature of space-time; and the right-hand side, which was the source of curvature-mass, but had to be arbitrarily put in, with no principle to determine the properties of mass. As in politics Einstein greatly preferred the left to the right. To quote Einstein: *What appears certain to me, however, is that, in the foundations of any consistent field theory the particle concept must not appear in addition to the field concept. The whole theory must be based solely on partial differential equations and their singularity-free solutions.*

So Einstein's goals were to:

(i) Generalize general relativity to include electromagnetism.
(ii) Eliminate the right-hand side of his equations and deduce the existence of matter by constructing singularity free solutions that would describe stable lumps of energy.
(iii) And finally, since he abhorred the arbitrary nature of the quantum rules and their probabilistic interpretation, he hoped to deduce them from these non-singular solutions.

He imagined that the demand of lack of singularities in the solutions that would describe matter would lead to over-determined equations, whose solutions would only exist for some, quantized values of physical parameters, say the radii of electron orbits. Thus he could imagine reproducing the Bohr model of the atom. The core of this program was to include electromagnetism and derive the existence of matter in the form of, what we call today, solitons. As Einstein understood, nonlinear equations can possess regular solutions that describe lumps of energy that do not dissipate. Thus one could start with the nonlinear field equations of general relativity and find localized particles. This was his hope:

> *'If one had the field equation of the total field, one would be compelled to demand that the particles themselves would everywhere be*

> *describable as singularity free solutions of the completed field equations. Only then would the general theory of relativity be a complete theory.'*

As far as I can tell, Einstein knew of no example of solitons or any toy model that exhibited his hopes. Nonetheless, flushed with the success of general relativity, with the faith that electromagnetism had to be unified, that matter needed a reason for its existence, he studied the equations and tried to modify them as well, with the hope of finding such solutions and with the dream that quantization of mass and charge, and even the quantum rules would emerge from overdetermination.

Among all of the extensions of general relativity considered and pursued by Einstein, the idea that the other forces of nature could be reflections of gravity in higher dimensions was the most innovative and enduring. It was not Einstein's idea, but rather that of Kaluza in 1922, significantly developed by Oscar Klein in 1926. Kaluza and Klein showed that if one assumed general relativity in five dimensions, where one dimension was curled up, the resulting theory would look like a four-dimensional theory of electromagnetism and gravity. Electromagnetism emerged as a consequence of gravity in five dimensions.

Einstein was immediately attracted to this idea and wrote to Kaluza — *'The idea of achieving (a unified field theory) by means of a five-dimensional cylinder world never dawned on me. At first glance I like your idea enormously.'* He held this paper for two years before submitting it to be published, probably because he was confused, as was Kaluza, as to whether the fifth dimension was real or not. Einstein returned again and again to this idea for over thirty years.

Einstein and Bergman in 1938 finally gave the best reasoning for taking the fifth dimension seriously, arguing that it is consistent with observation if it is sufficiently small. Klein had identified the momentum of particles moving around the fifth dimension as electric charge, which is quantized if one assumes the quantum mechanical rules of momentum quantization on circle. In modern versions of Kaluza–Klein, as they appear in string theory, this scenario is greatly amplified. In string theory there are six or seven extra-spatial dimensions. One can imagine that these are curled up to form a small manifold, and remarkably such six dimension compactifications (achieved by solving the generalization of Einstein's equations in ten dimensions) can produce a world remarkably like our own, in which the shape of the extra dimensions determines the complete matter content and all the forces of nature, as seen by a four-dimensional observer.

Why did not Einstein consider higher dimensional spaces? Much later he did play, for a while, with an eight-dimensional universe, a kind of complexification of Minkowski space, an approach severely criticized by Pauli and rapidly dropped by Einstein. But why did he not search systematically for higher dimension theories? If he had done so he might have discovered non-Abelian gauge theories, much as Oscar Klein almost did in 1938. I do not know, but suspect that part of the reason was that Einstein by and large ignored the nuclear forces altogether. His goal was to incorporate electromagnetism together with gravity — for this one extra dimension sufficed.

Einstein never thought much of this quantization of electric charge. Perhaps he thought, as Klein tried, to turn this around and derive the quantum rules from the quantization of charge. But in any case Einstein's main goal was to find particles as non-singular solutions of his equations and thus turned immediately to trying to find non-singular solutions of Kaluza–Klein theory.

Over the years Einstein came back again and again to this problem and tried to find non-singular solutions of Kaluza–Klein theory. He published at least three papers in which he proved that such solutions do not exist, with ever increasing generality. The last of these was a paper published with Pauli, who spent some of the war years in Princeton. The remark made in this paper that: *When one tries to find a unified theory of the gravitational and electromagnetic fields, he cannot help feeling that there is some truth in Kaluza's five-dimensional theory*, expressed how much Einstein was attracted to this approach. He must have been incredibly disappointed that he could not find matter as solitons in this theory.

But Einstein was wrong. There do exist solitons, non-singular solutions of his equations in Kaluza–Klein theory, which behave as particles — magnetic monopoles, with quantized magnetic charge. These were discovered in the early 1980s, by Perry and me, and independently by Sorkin, when Kaluza–Klein theory was revived. In our paper we added a footnote pointing out that these solutions contradicted Einstein. The referees suggested that we remove the footnote since it was disrespectful. We, of course, refused, how could we resist.

I have wondered what would have happened if these solutions had been discovered back in the 1920s; they could have. It would have given an enormous boost to Einstein's program, even though the solitons were magnetic and not electric, and very massive. But this did not happen and Einstein's attempts to find non-singular solutions failed, as did his attempts to construct satisfactory unified theories.

After sometime in the late 1920s Einstein became more and more isolated from the mainstream of fundamental physics. To a large extent this was due to his attitude towards quantum mechanics, the field to which he had made so many revolutionary contributions. Einstein, who understood better than most the implications of the emerging interpretations of quantum mechanics, could never accept it as a final theory of physics. He had no doubt that it worked, that it was a successful interim theory of physics, but he was convinced that it would be eventually replaced by a deeper, deterministic theory. His main hope in this regard seems to have been the hope that by demanding singularity free solutions of the nonlinear equations of general relativity one would get an overdetermined system of equations that would lead to quantization conditions.

Because of his opposition to quantum mechanics he allowed himself to ignore most of the important developments in fundamental physics for over twenty five years, as he himself admitted in 1954, *'I must seem like an ostrich who buries its head in the relativistic sand in order not to face the evil quanta.'* If there is one thing that I fault Einstein for, it is his lack of interest in the development of quantum field theory. To be sure many of the inventors of quantum field theory were soon to abandon it when faced with ultraviolet divergences, but it is hard to understand how Einstein, could not have been impressed with the successes of the marriage of his children quantum mechanics and special relativity. The Dirac equation and quantum electrodynamics had remarkable successes, especially the prediction of anti-particles. How could Einstein not have been impressed?

The only way to understand this is that general relativity was so important to him as to eclipse everything else. As Pauli remarked: *'If we would have presented Einstein with a synthesis of his general relativity and the quantum theory — then the discussion with him would have been considerably easier.'* But since general relativity and quantum mechanics seemed so incompatible, a situation that continued until quite recently, he felt free to ignore the exciting advances that were made in special relativistic quantum mechanics.

I turn now to the situation today, or more precisely thirty years ago, after the completion of the standard model of elementary particle physics, where we now have direct evidence for the unification of all forces dreamed by Einstein.

One of the most important implications of asymptotic freedom is the insight it gave into the unification of all the forces of nature. Almost immediately after the discovery of asymptotic freedom and the proposal of quantum chromodynamics, the first attempts were made to unify all the forces. This was natural, given that one was using very similar theories to describe all the known interactions. The apparently insurmountable barrier to unification, namely the large difference in the strength of the strong and the electroweak force, was seen to be a low energy phenomenon. Since the strong force decreases with increasing energy, all forces could have a common origin at very high energy. Indeed the couplings run in such a way as to merge about 10^{14} to 10^{16} Gev, close to the point where gravity becomes equally strong. This is our most direct clue as to where the next threshold of fundamental physics lies, and hints that at this immense energy all the forces of nature, including gravity, are unified.

In more recent times this extrapolation has greatly improved, due to the beautiful measurements of many experimenters and the hard work done by many theorists. Now the forces all meet only if we hypothesize a new space-time symmetry–supersymmetry — and if this new symmetry is broken at reasonably low energy; increasing hopes that a new super-world will be revealed at the Large Hadron Collider, soon to be completed at CERN. Supersymmetry is a beautiful, natural and unique extension of relativistic and general relativistic symmetries of nature. Einstein would, if he had studied it, have loved it. It can be thought of as the space-time symmetries of super-space, a space-time with extra dimensions. But the extra dimensions, here denoted collectively by θ, are measured with anti-commuting numbers. These are generalizations of ordinary real numbers, much as imaginary or complex numbers are; numbers that anti-commute, so that multiplication depends on the order, thus $\theta_1\theta_2 = -\theta_2\theta_1$. If it is hard to imagine a space of four or more dimensions, super-space is even weirder, but totally mathematically consistent. A theory formulated in super-space, and invariant under transformations or rotations of super-space, has many beautiful and appealing features. Supersymmetric extensions of the standard model can solve many important problems, such as why is there this enormous disparity, at low energy, between the strength of the gravitational force and the other forces of nature. The discovery of supersymmetry, which we all hope and some expect in a few years from now at the Large Hadron Collider, would be tantamount to the discovery of quantum dimensions of space-time.

Perhaps the most important feature of the extrapolation of the standard models forces is that the energy at which they appear to unify is very close, if not identical, to the point at which gravity becomes equally strong. This indicates that the next stage of unification should include, as Einstein expected, unification of the non-gravitational forces and gravity.

It is an important clue to that unification since it is not easy to quantize general relativity. A straightforward quantization of Einstein's theory does not work; the quantum fluctuations of the metric, at the characteristic distance scale of gravity, where the force becomes strong are too violent and uncontrollable. It seems inescapable that Einstein's theory is only an effective theory, adequate at long distances, but to be replaced by a more fundamental theory at the Planck scale of 10^{-33} cm.

Luckily such an extension of general relativity is available — string theory. String theory was not invented to describe gravity; instead it originated in an attempt to describe the strong interactions, wherein mesons can be thought of as open strings with quarks at their ends. The fact that the theory automatically described closed strings as well, and that closed strings invariably produced gravitons and gravity, and that the resulting quantum theory of gravity was finite and consistent is one of the most appealing aspects of this theory. String theory is a theory in development. We have learned much about this theory in the last decades, but much more remains. What has been achieved so far?

First, string theory is a consistent logical extension of the conceptual framework of fundamental physics. Such an extension is not easy and it is rare.

Second, string theory provides us for the first time with a consistent and finite quantum theory of gravity. This not only proves that quantum mechanics and general relativity are mutually compatible, it also provides us with the tools to explore many of the paradoxical issues that arise when the metric of space-time is quantized. Already string theory has clarified many of the mysteries of black holes. Thus the suspicion raised by Hawking as to whether black holes indicate the loss of information in fundamental physics has been dispelled, even to the point where Hawking himself has agreed that information is not lost in the process of formation and evaporation of black holes.

Finally, string theory has a rich structure that could yield a theory that unifies all of the forces of nature and explain all the constituents of matter. It automatically contains gravity as well as the gauge theories of the standard model. Certain of its four-dimensional compactifications give rise to low energy dynamics that is remarkably close to the standard model.

But string theory is still in the process of development, and although it has produced many surprises and lessons it still has not broken dramatically with the conceptual framework of relativistic quantum field theory. Many of us believe that ultimately string theory will give rise to a revolution in physics, as important as the two revolutions that took place in the 20th century, relativity and quantum mechanics. These revolutions are associated with two of the three fundamental dimensionful parameters of nature, the velocity of light and Planck's constant. The revolution in string theory presumably has to do with Newton's constant, that defines a length, the Planck length of 10^{-33} cm. String theory, I believe, will ultimately modify in a fundamental way our concepts at distances of order this length.

Where will the revolution take place? I believe that it will involve our understanding of the nature of space-time, a subject dear to Einstein's heart. To quote some leading string theorists:

Space and time may be doomed. — E. Witten
I am almost certain that space and time are illusions. — N. Seiberg
The notion of space-time is clearly something we're going to have to give up. — A. Strominger
The real change that's around the corner is in the way we think about space and time. We haven't come to grips with what Einstein taught us. But that's coming. And that will make the world around us stranger than any of us can imagine. — D. Gross

Why is space-time doomed? There are many reasons, among which: In string theory we can change the dimension of space-time by changing the strength of the string force. Thus, the so-called II-A string theory, which semi-classically describes closed strings moving in ten-dimensional flat space for very weak coupling is dual for strong coupling to a theory, called M-theory, that at low energies is described by eleven-dimensional supergravity. By increasing the string coupling we can grow an extra dimension. How can the spatial continuum be fundamental if the number of spatial dimensions can be so changed?

We can continuously tear the fabric of space. Thus a string theory solution that describes strings moving on a background wherein some of the spatial dimensions are compactified on a manifold M_1 can be continuously deformed, by varying some of the parameters of the solution, to one that describes the strings moving on a background M_2 of different topology. In between there is no such simple description of the solution as strings moving on a geometric background, but the deformation is continuous and the strings do not mind at all that the fabric of space has been torn so as to modify the topology. Again this suggests that the spatial continuum cannot be fundamental if its topology can be changed in this smooth fashion.

On the other hand in string theory we cannot probe arbitrarily small distances. In string theory we can ask what is the smallest distance that can operationally be explored, analyzing (as Heisenberg did in the case of quantum mechanics) how a microscope works. In string theory the light rays of a microscope are really strings. Consequently, as we increase the energy of the light, so as to overcome the quantum mechanical uncertainty in the measurement of distance, the strings expand and prevent us from resolving arbitrarily small distances. The minimum distance that we can explore is, not surprisingly, of order the Planck length.

We also cannot squeeze spatial volumes to zero size. If one of the spatial dimensions is compactified to form a circle of radius R, it turns out that string theory in this background is identical to string theory in a background where the radius of this circle is $1/R$ (in Planckian units). Thus if we try to squeeze this dimension and reduce R to zero, we find that the more natural description is in terms of the dual theory, and the minimal size of the compact circle is finite and of order the Planck length.

These phenomena suggest that there is no operational meaning to distances smaller than the Planck length, that the spatial continuum should be replaced by something else. I believe that space for sure, and presumably time as well, will be emergent. We already have many hints and examples where space is an emergent concept. These include the famous AdS/CFT duality, wherein string theory in ten dimensions, with a background geometry of five-dimensional Anti-DeSitter space times a five sphere, is dual to supersymmetric gauge in flat four-dimensional space-time. Six spatial dimensions emerge from the gauge theory description, together with gravity. We have no understanding, however, what it would mean that time itself would be an emergent concept.

I like to depict our confusion in poetic form. Democritus expressed 2500 years ago the atomic hypothesis in the following verse:

By convention there is color,
by convention sweetness,
by convention bitterness,
But in reality there are atoms and space.

I say: We are convinced that

By convention there is space,
By convention there is time,
But in reality there is...

The problem is that I do not know how to finish the verse.

So did Einstein go wrong in the latter part of his life? The answer is both yes and no.

Yes, he refused to accept quantum mechanics. He ignored the developments in nuclear and particle physics. These mistakes ensured his failure, but they are quite understandable and forgivable.

No, he knew that gravity must be unified with the other forces. And this we too know today is the central issue in fundamental physics.

And for those of us faced with the fact that we cannot yet directly probe the Planck scale, he believed in the possibility of successful speculative theory. As Einstein stated: *'The successful attempt to derive delicate laws of nature, along a purely mental path, by following a belief in the formal unity of the structure of reality, encourages continuation in this speculative direction, the dangers of which everyone vividly must keep in sight who dares follow it.'*

To all physicists, but especially to those working in speculative areas, Einstein remains an inspiration for his foresight, and his unyielding determination and courage.

CHAPTER 2

Einstein and Geometry

****

MICHAEL ATIYAH

School of Mathematics, University of Edinburgh,
Edinburgh EH9, 3JZ, UK
m.atiyah@ed.ac.uk

Einstein initiated and stressed the role of geometry in fundamental physics. Fifty years after his death the links between geometry and physics have been significantly extended with benefits to both sides.

1. General Relativity

Einstein is generally recognized as the greatest physicist of the 20th century and perhaps the greatest physicist since Newton, though Faraday and Clerk Maxwell are close competitors. Einstein is a case where popular acclaim and scientific standing are in agreement. But unlike Newton, Einstein was not a mathematician. He used mathematics in an essential way but he did not create it and he relied on his colleagues for technical help. It is all the more remarkable that his ideas have triggered great advances in geometry, even in parts of the subject apparently far removed from physics.

I will attempt to describe and explain how this has come about. But first I should make some general remarks about the relation between physics and mathematics. The conventional view is that mathematicians have developed machinery for studying numbers (which might represent physical quantities) and the way in which those relate to each other in the form of equations. Physicists then use this language and embody their conclusions in 'laws' described by equations. Thus Newton's gravitational theory is described by the inverse square law of mutual attraction, while the fundamental laws of electromagnetism are encoded in Maxwell's equations.

While this orthodox view is formally correct, it hides some essential features. In physics the starting points are the concepts: particles, forces, space, time, motion, interaction. Objects are seen to move around and act on one another. The secondary part of the story is the taking of measurements by the experimental scientist. Numbers are written down, tabulated, compared.

The earliest part of mathematics to be studied in depth was geometry, in the hands of the Greeks. The basic concepts here are: points, lines, angles, triangles, circles and their mutual relation. Numbers, giving distances and areas come shortly thereafter, but equations did not enter the picture until the work of Descartes in the 17th century.

The connection between physics and geometry starts at the conceptual stage in a fully three-dimensional picture of the world, and has nothing to do with any reference frame in which one may choose to take measurements. It is not easy to move from physics to geometry without choosing (x, y, z) coordinates and writing equations, but it is more fundamental. Descartes' introduction of coordinates may have been an essential step in the formalization of mathematical physics but it was also an abdication: it gave up on trying to understand physics geometrically.

Newton understood this, which is why he presented his *Principia* in geometric form, but this was too difficult for posterity which followed the ideas of Descartes and Leibniz.

This brief philosophical review is essential if we want to understand how Einstein's ideas came to influence geometry. As we all know, Einstein's monumental contribution was the replacement of the Newtonian theory of gravity by what is called General Relativity. This theory has two essential features, the first is to move from three-dimensional geometry to four-dimensional geometry by incorporating time as a fourth variable. This is the content of Special Relativity, but the second key step is to interpret gravitation as the curvature of this four-dimensional space-time geometry.

Standard textbooks make great play with the technical details, introducing coordinates, writing equations and then showing that the resulting physics is independent of the choice of coordinates. To a geometer this is perverse. The fundamental link is from physics to geometry, from force to curvature and the algebraic machinery that encodes this is secondary. God created the universe without writing down equations!

2. Electromagnetism

As far as gravity is concerned, Einstein's General Relativity is a beautiful and complete theory. But as Einstein realized it has to be extended to account for other physical forces, the most notable being electromagnetism. It is perhaps no accident that the first and most significant step in this direction was taken by a mathematician — Hermann Weyl. He showed that, by adding a fifth dimension, electromagnetism could also be interpreted as curvature. His idea was that the size of a particle could alter as it passed through an electromagnetic field. In analogy with railways it was called a gauge theory, and this name has stuck through subsequent evolutions of the theory.

Unfortunately for Weyl, Einstein immediately objected on physical grounds that this would have meant different atoms of, say hydrogen, would have different sizes depending on their past history, in contradiction with observation. Given this devastating critique, it is remarkable but fortunate that Weyl's paper was still published, with Einstein's objection as an appendix. Clearly the beauty of the idea attracted the editor, despite the fatal flaw. In fact, beauty often wins such contests, because with the advent of quantum mechanics, with its complex wave functions, it was pointed out by Kaluza and Klein that Weyl's gauge theory could be salvaged if one interpreted the variable as a phase rather than a length. A pure phase shift by itself is not physically observable and so Weyl's theory avoids the Einstein objection.

Quantum mechanics

While quantum mechanics thus came to the rescue of Weyl's gauge theory and so continued the Einstein programme of geometrizing physics, it also seemed to demolish the whole idea. While quantum mechanics is a very subtle and beautiful mathematical theory, it strays very far from geometry and is conceptually difficult to comprehend. In fact, as is well known, Einstein never fully accepted quantum mechanics as the final word. He disliked its philosophical basis with its need for probability and uncertainty.

While conceding its great practical success, Einstein remained opposed to quantum mechanics to his dying day. Increasingly he was regarded by the younger generation of physicists as being obstinate and out of touch. His continued search for a unified field theory only confirmed this widely held opinion.

3. Nuclear Forces

Einstein and Weyl, who both went to the Institute for Advanced Study in Princeton as refugees from Germany in the 1930s, died in 1955, the year

I myself went to Princeton as a fresh Ph.D. This was also the year when Yang–Mills theory was born, the theory which developed in due course into the standard framework for understanding the 'weak' and 'strong' forces which operate on the nuclear scale and are believed, together with gravitation and electromagnetism, to provide all the fundamental forces of nature.

Yang–Mills theory can be roughly understood as the natural extension of Maxwell's theory in which the angular phase is replaced by a phase specified by rotation in a higher dimensional 'internal space'. This internal space is not part of our usual space-time but is additional to it, just as the Maxwell phase was interpreted as a fifth dimension. There is one fundamental difference between angles (rotation in a plane) and rotations in three or more dimensions. Two such rotations, about different axes, do not in general 'commute', that is to say that the result of performing the two rotations one after the other depends on the order in which they are performed. This is easy to verify by considering rotations of the earth. Consider for example, a rotation A about the North Pole/South Pole axis of say 20° in a westward direction, and a rotation B around the axis through Chennai (and its antipode), which takes Bangkok to a position due North of Chennai (somewhere in Northern Kashmir). Performing first B and then A will take Bangkok, via Kashmir, to northern Iran. On the other hand, performing A first will take Bangkok approximately to Chennai, so that following this by B will leave it there. The results obviously differ — Chennai is not in Iran!

This non-commutativity of rotations has major consequences for Yang–Mills theory, making it a much more complicated and subtle theory than Maxwell theory. In particular it becomes nonlinear, which has profound mathematical and physical consequences.

It is somewhat ironic that the ideas of Yang and Mills developed quite independently of Weyl and Einstein and that there was little interaction with them. No doubt the generation gap was too large. In addition, Yang–Mills theory was a quantum theory, still in its infancy, and the full geometrical implications were not yet apparent.

With the belated recognition that all four of the fundamental forces of nature were geometrical, one might have said that Einstein's dream of a unified field theory was finally realized, even if it came after Einstein's death. In fact, as has just been indicated, this was only partially true because of the presence of quantum theory. On the one hand the quantum aspects made the theory extremely difficult and sophisticated, taking further decades to unravel. On the other hand, Einstein's philosophical objections would remain. He would still be dissatisfied from beyond the grave. Nevertheless physicists now grudgingly acknowledge that Einstein's intuition was in part justified and that the revolution he introduced in General Relativity of geometrizing physics has proceeded much further. Perhaps the verdict would be that the final outcome of the long Einstein–Bohr arguments was a draw, with the big proviso that 'finality' has not yet been achieved.

4. String Theory

At this point in the development, although geometry provided a common framework for all the forces, there was still no way to complete the unification by combining quantum theory and general relativity. Since quantum theory deals with the very small and general relativity with the very large, many physicists feel that, for all practical purposes, there is no need to attempt such an ultimate unification. Others however disagree, arguing that physicists should never give up on this ultimate search, and for these the hunt for this final unification is the 'holy grail'.

In the past thirty years a promising framework has appeared in which such a unification seems conceptually possible. This is 'string theory', based on the simple idea that point particles should be replaced by one-dimensional objects — strings, either open (with free ends) or closed (in circular form).

String theory, which is as yet unfinished and incomplete, involves yet more geometry beyond Yang–Mills. In the first place a string moving in time spans a surface which has its own geometry. For example, the surface may acquire holes, a topological feature with profound implications, already known in mathematics. In the second place consistency of the physical theory requires that the string should be not just in ordinary three-dimensional space, but in one of nine dimensions (or ten if one includes time). Both of these open up vast new (geometrical) territories which strengthen the link between geometry and physics. This works both ways, first large amounts of mathematics developed over previous centuries suddenly become relevant and available for physicists to use. Second, and perhaps more surprising, the ideas of physics including quantum field theory feed back into mathematics and lead to surprising developments. In fact the mathematical activity generated by this interaction with physics is, in my opinion, the most exciting development in mathematics of the past decades, and we seem still to be in the early stage.

5. M-Theory

After the initial rapid development of string theory, a drawback appeared when it was realized that there were five different competing models of string theory. There seemed no reason why nature should prefer one to another.

But it was eventually discovered that all these five string theories were, in a subtle way, equivalent to each other. The best analogy is provided by the basic calculus of analytic functions, in which a function $f(z)$ can be expanded as a power series in z. As a simple example, the binomial theorem tells us that

$$(1 + z)^3 = 1 + 3z + 3z^2 + z^3 .$$

However if we introduce a variable $u = 1 - z$, then

$$(2 - u)^3 = 8 - 12u + 6u^2 - u^3 ,$$

represents the same function, so the two polynomials are really equivalent under a simple change of variable. String theories are like such power series expansions, but the equivalence between two of them is much subtler than a change of variable.

The five different string theories are now seen as different viewpoints of one underlying theory, which is not yet known but has been christened 'M-theory'. By analogy you might be given the power series expansion of say $\sqrt{(1 - z)}$ about several different values of z and you might be able to check that they were equivalent without recognizing that the function was a simple square root.

While a proper understanding of M-theory still eludes us, much is now known about it. In particular, the various geometric results that have emerged from string theory become related in interesting but mysterious 'dualities' whose real meaning has yet to be discovered.

No one can predict what the future holds in store for M-theory. Are we nearly there, is the final understanding just round the corner? Will it come from a few more technical tricks or will it require some fundamental breakthrough? The biggest question of all and the one that Einstein would still be asking is: can M-theory be properly understood within the present framework of quantum mechanics or do we need to look for new foundations? I confess that I myself remain an Einsteinian and would be happy to see quantum mechanics replaced by something deeper. This remains, as in Einstein's day, a minority opinion but one shared, for example, by Roger Penrose.

6. Topology

I have alluded at various stages to the impact that these physical theories have had on geometry, without providing much detail. Let me now try to rectify this.

Classical physics, describing various forces, is closely linked (via Einstein and Maxwell) with notions of curvature in geometry. The connection between physics and geometry is therefore local: we can study the forces in a small piece of space-time and compare it with the local geometry. By contrast quantum physics is not related to geometry in this way. Its relation is 'global' and can only be seen in the whole picture (even if we are dealing with microscopic objects). The global aspect of geometry that is involved is 'Topology', such as the study of holes in surfaces or of knots in three-dimensional space.

The first indication that quantum mechanics was related to topology was in the argument of Dirac which explained why the electric charge of any particle was an integer multiple of the charge of the electron. The integers came in essentially as 'winding numbers', counting the number of circuits made by a closed path. This number is topological because it does not depend on the detailed local geometry of the circuit, how long or wiggly it is, but only its overall global behavior.

Winding numbers are related to the circle and hence to the angular phase of electromagnetism. There are similar but more complicated topological properties associated with the higher dimensional phases of Yang–Mills theory so that the relation between quantum theory and topology carries over to the other forces. In addition, as pointed out earlier, moving strings generate surfaces which may have holes and these are topological in nature.

So the physics of string theory and M-theory is replete with topological information and many intricate and subtle aspects of the quantum theory are related to this underlying topology.

So what kinds of specific geometrical/topological results have emerged from the interaction with physics? In fact these are quite diverse and cover many types of problems. Here is a short list.

Knot invariants

The study of knots (closed pieces of string) is a standard but difficult branch of topology. The key problem is to find 'invariants' which will distinguish essentially different knots. An invariant is something (a set of numbers) which can be calculated from a picture of the knot, but is unaltered if we move the knot around to get a different picture. In the 1970's the world of topologists was astounded when the New Zealand mathematician, Vaughan Jones discovered a new type of invariant which helped to solve 100-year old problems. Shortly after, Edward Witten gave a physical explanation of the Jones invariants which cast new light on them and led to much further progress.

Donaldson invariants

Geometers have studied the topology of closed surfaces and their higher-dimensional analogues (manifolds) for a long time. But a remarkable breakthrough came in the early 1980s when Simon Donaldson found some totally new and unexpected invariants of four-dimensional manifolds. These were based on the Yang–Mills equations of physics but it was not until later that Edward Witten again showed how to interpret Donaldson's invariants in terms of quantum field theory. Later still, using duality ideas from string theory, Witten and Seiberg made a significant improvement of Donaldson theory which led to solutions of old problems.

Counting curves

Classical algebraic geometers, ever since the time of Descartes, studied curves in the plane given by polynomial equations. It is a natural question to ask how many curves there are of a given type, passing through a given number of points. For example, there is a unique straight line through any two points and a unique conic (ellipse, etc.) through five points. The question gets harder as the curves get more complicated and given by polynomials of higher degree. Quite remarkably, ideas from string theory have led to a complete solution of this problem.

These and other examples are now part of a broad area of 'quantum mathematics' — an evocative term which correctly conveys the origin of the ideas and results but is very loosely used and ill-defined. One of the big challenges for mathematicians at the present time is to see if one can understand these new mathematical theories without recourse to the physical background. Alternatively, it may become necessary to incorporate or absorb various physical ideas into rigorous mathematics.

The converse process of providing rigorous mathematical treatment of quantum field theory, string theory, M-theory appears a very distant prospect. It will certainly have to wait till physicists have sorted themselves out and allowed the dust to clear.

7. Conclusion

Einstein would, I think, have been both surprised and gratified by the extent to which his geometrization of physics has progressed. The mathematical by-products would have surprised him even further. But the fact that his ideas were so fruitful would only encourage him in his fundamental beliefs. In particular, he would still be encouraging us to dig beneath the mysteries of quantum mechanics. In another century we might find what Einstein was looking for.

CHAPTER 3

String Theory and Einstein's Dream

****

ASHOKE SEN

Harish-Chandra Research Institute,
Chhatnag Road, Jhusi, Allahabad 211019, India
ashoke.sen@cern.ch
sen@mri.ernet.in

Unification of the theory of gravitation, as given by Einstein's general theory of relativity, and the theory of electromagnetism, as formulated by Maxwell, had been Einstein's dream during the later part of his life. String theory, which is the subject of this article, is an attempt to realize this dream. However in many ways, string theory attempts to go beyond Einstein's dream. String theory attempts to bring all known forces of nature — not just gravity and electromagnetism — under one umbrella. It also tries to do so in a manner that is consistent with the principles of quantum mechanics — the theory that is necessary for describing the laws of nature at very small distance. Thus, string theory is an attempt to provide an all-encompassing description of nature that works at large distances where gravity becomes important as well as small distances where quantum mechanics is important.

1. Introduction

Unification of the theory of gravitation, as given by Einstein's general theory of relativity, and the theory of electromagnetism, as formulated by Maxwell, had been Einstein's dream during the later part of his life. String theory, which is the subject of this article, is an attempt to realize this dream. However, in many ways string theory attempts to go beyond Einstein's dream. String theory attempts to bring all known forces of nature — not just gravity and electromagnetism — under one umbrella. It also tries to do so in a manner that is consistent with the principles of quantum mechanics — the theory that is necessary for describing the laws of nature at very small distance. Thus string theory is an attempt to provide an

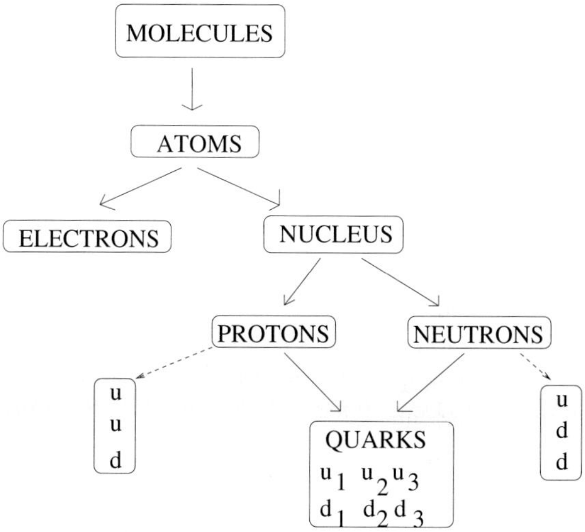

Fig. 1. Our current understanding of the building blocks of matter.

all encompassing description of nature that works at large distances where gravity becomes important as well as small distances where quantum mechanics is important.

In this article, I shall try to give a very general introduction to string theory.[1] However, in order to do so, I must begin by reviewing our current understanding of the basic constituents of matter. This is the subject to which we shall now turn.

2. The World of Elementary Particles

According to our current understanding, everything that we see around us is made of a few elementary building blocks. Figure 1 gives us a bird's eye view of our current knowledge of the structure of matter. At the crudest level the building blocks of matter are the individual molecules of various compounds. However, there are a very large number of compounds, each with its own characteristic molecule. A simpler picture emerged when it was realized that each molecule is made of some smaller building blocks known as atoms. There are about 100 different types of atoms and different molecules differ in their properties because they contain different

[1] Refs. [1–4] provide some good introductory textbooks on string theory.

number of atoms of different types in different arrangements. During the early years of the twentieth century it was realized that atoms are also not the smallest constituents of matter — each atom is made of a central nucleus and a set of electrons revolving around it. Different atoms have different number of electrons, but all the electrons found in all atoms have identical properties. In contrast the nuclei of different types of atoms have very different properties. This picture simplified once it was realized that each nucleus can be regarded as being made of even smaller constituents — the proton and the neutron. Different nuclei have different properties because they contain different numbers of protons and neutrons. Finally, even the protons and neutrons are now known to be made of even smaller constituents called quarks — the proton being made of two up (u) quarks and one down (d) quarks, and the neutron of one u and two d quarks. According to our current knowledge, the electrons and the quarks cannot be divided any further. We call them elementary particles.

This gives us a very simple picture of the structure of matter, namely everything is made of three different types of 'elementary particles' — the electron, the u quark and the d quark. However, as we shall see, this is far from a complete picture. As is already evident from Fig. 1, the up and down quarks each come in three varieties. Here, we have denoted them by u_1, u_2, u_3 and d_1, d_2, d_3, but often they are referred to as red, blue and green type of quarks. We shall refer to this as the color quantum number although this has nothing to do with the color that we see in everyday life. The quarks inside the proton and neutron continuously change their color due to a process known as strong interaction that will be discussed soon. There are various other reasons why this picture is not complete. I shall review some of them here.

In order to understand the structure of matter, we need to understand not only the basic constituents of matter, but also the nature of the forces that operate between them. Without this knowledge we shall not have any understanding of what keeps the quarks bound inside a proton and neutron, or at a larger scale, of what keeps the atoms bound inside a molecule. According to our current knowledge there are four basic types of forces operating between elementary particles — (1) gravitational, (2) electromagnetic, (3) strong and (4) weak. Of these the gravitational and the electromagnetic forces are familiar to us from everyday experience. For example, the gravitational force is responsible for earth's gravity and the motion of the planets around the sun. The electromagnetic force is the cause of lightening in the sky, the force of a magnet, the working of

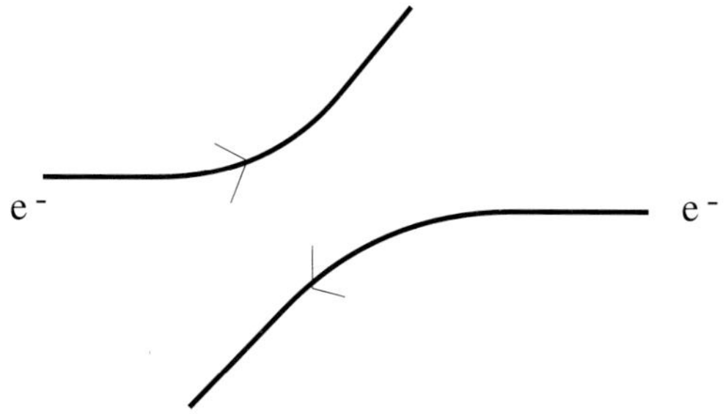

Fig. 2. Classical picture of the deflection of a pair of electrons via electromagnetic force.

various electrical appliances, etc. It is also responsible for binding the electrons and the nuclei inside the atom and the atoms inside a molecule. The strong force operates between quarks and is responsible for binding them inside a proton and a neutron and also for binding the proton and the neutron inside a nucleus. The weak force, being weak, is not responsible for binding any particles; however it is responsible for certain radioactive decays known as β-decay.

It turns out that in studying the physics of elementary particles, we can ignore the effect of gravitational force. To see this one can compare the electrostatic force between two protons with the gravitational force between two protons at rest. The result is

$$\frac{\text{Grav. Force}}{\text{Elec. Force}} = \frac{G_N m_p^2/r^2}{e_p^2/r^2} \sim 10^{-36}$$

where G_N is the Newton's constant (6.67×10^{-8} cm^3/gm sec^2) that controls the strength of the gravitational force between two bodies, m_p is the proton mass (1.67×10^{-24} gm) and e_p is the proton charge (4.8×10^{-10} e.s.u.). Clearly this ratio is extremely small. Similarly all other forces can also be shown to be much larger than the gravitational force.

So far we have discussed the elementary particles and the forces operating between them as separate entities, but with the help of quantum theory one can give a unified description of elementary particles, and the forces among the elementary particles. Consider for example the electromagnetic force between two electrons when they pass each other. Due to this force, each particle gets deflected from its original trajectory. This has been depicted in Fig. 2. In quantum theory, one provides a different explanation of the same phenomenon. Here the deflection takes place because the two electrons exchange a new particle, called photon, while passing near each

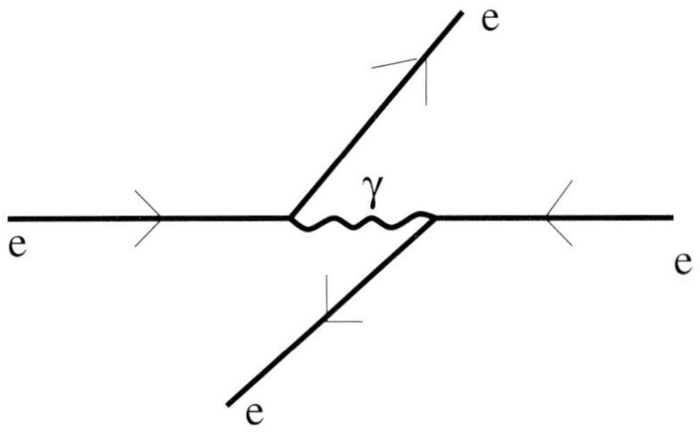

Fig. 3. Quantum picture of the deflection of a pair of electrons via electromagnetic force.

other (see Fig. 3). The photon is capable of carrying some amount of energy and momentum from the first electron to the second electron, thereby causing this deflection.[2] We call the photon the mediator of electromagnetic force. Even though it mediates electromagnetic force, the photon itself is electrically neutral.

Thus in the language of quantum theory we can describe a force by specifying the particle(s) which mediate the force. It turns out that the strong force is mediated by eight different particles known as gluons. These particles are all electrically neutral. The quarks inside a proton (and neutron) continuously exchange gluons, and in this process keep changing their color quantum number. On the other hand, the weak force is mediated by three particles, denoted by W^+, W^- and Z. W^+ and W^- carry $+1$ and -1 unit of electric charge respectively while Z is neutral. (The unit of electric charge is taken to be the charge carried by a single proton. Thus W^+ has charge equal to that of a proton, while W^- has charge that is equal in magnitude but opposite in sign to that of a proton.)

Clearly, we must add the gluons, W^+, W^- and Z, as well as the photon, to our list of elementary particles. We shall refer to these as the mediator particles. Theoretical analysis shows that for every elementary particle there must also be another elementary particle, known as the antiparticle, that carries exactly the same amount of charge but with opposite sign. Thus for every quark and the electron we have the corresponding anti-quark

[2]The quantum picture shown in Fig. 3 suggests that the change in the direction of the electrons happens suddenly instead of continuously. In practice, each exchange of photon causes a tiny amount of sudden jump, and the classical picture emerges due to the quantum process repeated many times via many exchanges of photons.

and the anti-electron (known as the positron). Fortunately the gluons, the photon and the Z particles are their own anti-particles, whereas W^- is the anti-particle of W^+ and vice-versa. Thus we do not need to expand our list by including anti-particles of the mediator particles. However, this still does not exhaust the list of all elementary particles. Besides the u and d quarks, electrons and mediators and their anti-particles, there are also other elementary particles which are produced by cosmic rays, radioactive decays, collision of high energy particles, etc. They must also be added to the list.

Our current list contains about 100 types of elementary particles. Thus the situation would not seem any better than the days when atoms were thought to be the basic constituents of matter. The properties of matter known at that time could be explained in terms of the properties of about 100 types of atoms. There is however a difference — unlike the case of atoms, there is a simple mathematical theory that explains the properties of all the elementary particles. In fact, this theory has been so successful that it has come to be known as the 'standard model' of elementary particles. This model, in principle, can be used to calculate the result of any experiment that we wish to perform involving the elementary particles. So far the standard model has been extremely successful in explaining almost all experimental results.

3. The Standard Model: Its Successes and Limitations

In this section I shall explain some of the basic properties of the standard model. The basic inputs in this theory are

- quantum mechanics,
- special theory of relativity, and
- laws of electromagnetism and their generalization to strong and weak forces.

There is a mathematical framework, known as gauge theory, that includes all these three features. I shall not describe the details of this framework here. It turns out that there are many different consistent gauge theories, one of which describes the theory of elementary particles. This particular theory is known as the standard model.

Once the theory is written down, it predicts the outcome of every possible experiment involving elementary particles. (Of course, some experimental inputs go in to decide on what is the right theory.) For example, the standard model tells us precisely what kind of elementary particles we have in our world. According to this model, the elementary particles in our world fall into four categories:

- **Quarks** $u_1, u_2, u_3,\ d_1, d_2, d_3,\ c_1, c_2, c_3,\ s_1, s_2, s_3,\ t_1, t_2, t_3,\ b_1, b_2, b_3$
  In this list we recognize the familiar up and down quarks, each coming in three colors. It turns out that nature contains four more types of quarks — charm (c), strange (s), top (t) and bottom (b), each coming in three colors. These four types of quarks are not usually found inside matter but can be produced in highly energetic collision among normal matter. Of the six quarks, the up, charm and top quarks carry 2/3 unit of electric charge, whereas the down, strange and bottom quarks carry $-1/3$ unit of electric charge. For each quark we also have its anti-quark; we have not listed them separately here.
- **Leptons** $e^-, \nu_e,\quad \mu^-, \nu_\mu,\quad \tau^-, \nu_\tau$
  In this list we recognize the electron (e^-); the $-$ sign on top is to remind ourselves that the electron carries -1 unit of charge, i.e. charge equal in magnitude but opposite in sign to that carried by the proton. ν_e — known as the electron neutrino — is a weakly interacting chargeless particle. These are so weakly interacting that a neutrino passing through the earth does so experiencing almost no force. The pair of particles (μ^-, ν_μ) have properties similar to that of the pair (e^-, ν_e) although the muon (μ^-) is a lot heavier that the electron. Similarly the pair (τ^-, ν_τ) have properties similar to that of (e^-, ν_e), with the tau particle (τ^-) being even heavier than a muon. For each lepton we also have an anti-lepton which we have not listed here. For example, the anti-particle of the electron is called the positron and denoted by the symbol e^+.
- **Gauge Bosons** gluons: $g_1, \ldots g_8$, Photon: γ, W^+, W^-, Z
  These are the by now familiar mediator particles which have been discussed before. As already mentioned the list is complete without having to add the anti-particles separately.
- **Higgs Particle** ϕ
  This is the most mysterious particle in the standard model. Unlike every other particle in the list which has been experimentally observed, the Higgs particle has never been seen in any experiment despite several attempts. Nevertheless its existence is predicted by the standard model, and new experiments are being designed to look for this particle.

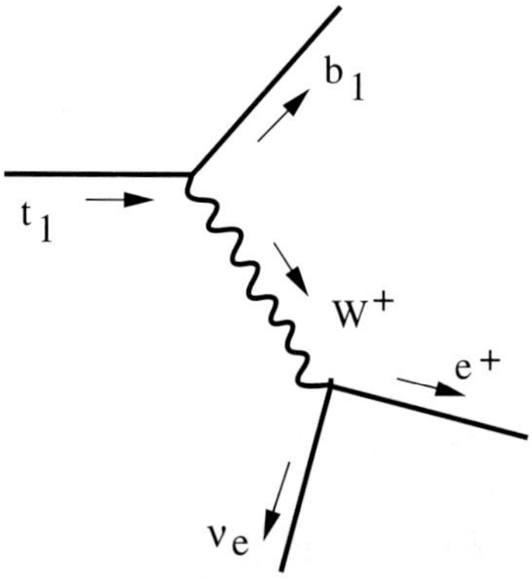

Fig. 4. An allowed process in the standard model.

The standard model not only gives us a list of elementary particles but also the list of processes that can occur involving these particles. For example, in order to explain the electromagnetic force between electrons using the process described in Fig. 3, it is necessary to know that an electron can emit a photon. This follows from the mathematical framework that lies behind the standard model. The same mathematical framework also tells us that if in this diagram we replace the electron by an electron neutrino then this is not an allowed process in the standard model; hence a neutrino cannot exchange a photon with another particle. Figure 4 shows another example of a process that can occur in the standard model. This describes the decay of a top quark (t_1) into an electron neutrino (ν_e), a positron (e^+) and a bottom quark (b_1). In fact the standard model not only tells us which processes can occur, but it also gives us precise mathematical formula for calculating the probablility of occurence of any such process. These predictions are then compared with experimental data to test the model.

Given the success of the standard model, one might like to conclude that we now have a complete understanding of the elementary constituents of matter. This however is not true. There are several reasons why standard model cannot be the complete theory of elementary particles. I shall review a few of these here.

First and foremost, the standard model does not explain the origin of one of the important forces that we observe in nature, namely the gravitational force. In particular, the list of particles predicted by the standard model does not contain any particle that mediates gravitational force. The effect of this omission, of course, is not seen in any of the experiments involving elementary particles since, as observed earlier in this article, the gravitational force between two elementary particles is extremely small compared to the other forces. Nevertheless, a complete theory must account for every possible tiny effect that exists in nature. Thus, a theory that does not provide an explanation of the gravitational force cannot be a complete theory of nature.

In order to appreciate the gravity of this problem, let us first take stock of what is known about gravity. Our current theoretical understanding of the gravitational force is based on the 'general theory of relativity' — a theory written down by Einstein almost a hundred years ago. This theory has been enormously successful in explaining all effects related to gravity. Unfortunately, this theory is based on the principles of classical mechanics and not of quantum mechanics. Since other forces in nature follow the rules of quantum mechanics, any theory that attempts to explain gravity as well as the other forces of nature must treat gravity according to the rules of quantum mechanics. Hence the general theory of relativity, despite being so successful, cannot be the final story about gravity. In fact, the reason that this theory has been so successful so far is that for gravity the difference between the predictions of a classical and the quantum theory is extremely tiny and cannot be observed in any of the current experiments. (We say that quantum effects involving gravity are extremely small.)

Thus the problem at this stage seems to be to first find a quantum theory of gravity and then combine this with the standard model to arrive at a complete theory of all elementary particles and forces operating between them. At the first sight the problem does not seem unsurmountable. After all, we normally obtain a quantum theory by first writing down a classical theory and then applying a definite set of rules to turn it into a quantum theory. Why cannot the same thing be done with the general theory of relativity? If one proceeds to do this one does get some encouraging results at first. In particular one finds that like other forces, gravity is also mediated by a new kind of elementary particle. This particle has been given the name graviton. Like the diagram in Fig. 3 one will have a diagram where two electrons exchange a graviton, representing the (tiny amount of) deflection of one of the electrons due to the gravitational force of the other electron.

So far everything seems to be proceeding as desired. However, one soon runs into a problem with this approach. To understand the origin of this difficulty consider the process shown in Fig. 5 involving multiple graviton exchanges. As in the case of the standard model, there are precise

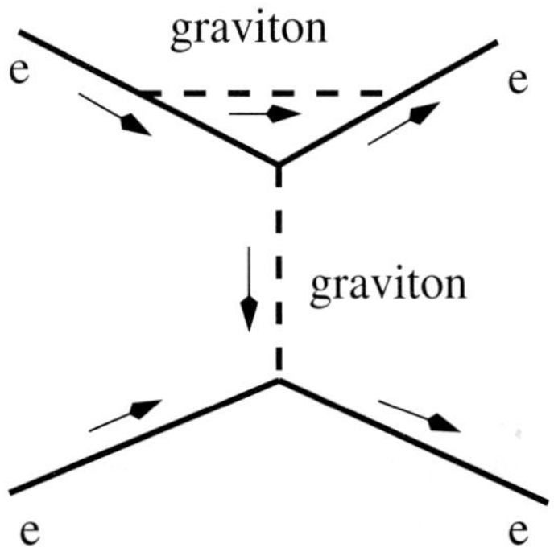

Fig. 5. An infinite contribution to the gravitational scattering of two electrons.

mathematical rules for computing the probability amplitude of this process in the quantum general theory of relativity. When one applies those rules to calculate this probability amplitude, one finds that the result is infinity!

This is clearly a nonsensical answer! In actual practice we know that this probability must be extremely tiny since no experiment has yet seen the effect of gravitational force between elementary particles. Thus there must be something wrong with this theory.

In order to appreciate how string theory eventually resolves this problem, it will be useful to investigate in a little more detail the origin of this problem. You would notice that in a diagram like the one shown in Fig. 5 there are 'interaction vertices' where three (or more) lines meet. For example in Fig. 5 there are four such interaction vertices. These are the points where something happens. We can regard these points as the basic events which make up the complete process. Each such event takes place at a given point in space at a given time, and in order to calculate the total probability amplitude of the process we must integrate over the location of each event in space as well as in time. It turns out that the integrand, calculated using the rules of quantum theory, diverges (becomes infinite) when more than two or more such elementary events take place at the same point in space at the same time. This in turn causes the integral to diverge occasionally.[3]

[3]Similar divergences also occur in the standard model, but can be removed by a procedure known as renormalization. This procedure does not work for general theory of relativity since the divergences are more severe.

In any case the final outcome of this complicated analysis is that the standard procedure that has been successful in formulating a quantum theory of strong, weak and electromagnetic forces do not work for gravity, and for this reason it is not easy to incorporate gravity into the standard model.

Besides the problem of incorporating gravity, the standard model suffers from other conceptual and technical problems. While it is true that the standard model, once formulated, can predict the results of most experiments involving elementary particles, the formulation of the theory itself requires a lot of input from experiments. For example, there are many consistent gauge theories, often labeled by several continuous parameters, and standard model corresponds to one of these theories with a specific choice of the values of these parameters. There is no explanation within the theory as to why this particular gauge theory with this particular choice of parameters should describe our universe. Furthermore the choice of parameters which describes the standard model are not generic, but requires very fine tuning. This is evident from the fact that the theory has some extremely small dimensionless numbers like the ratio of gravitational and electromagnetic force between two elementary particles. For a generic choice of parameters this ratio would be of order one. Finally recent experiments show that not all predictions of the standard model are completely correct. In particular, according to the standard model the neutrinos are zero mass particles, but recent experiments show that neutrinos actually have a tiny but finite mass. This requires a small modification of the gauge theory that describes the standard model.

These are some of the reasons why we believe that the standard model is not the final story. In the rest of this article we shall try to see how string theory attempts to address some of these issues.

4. String Theory

The basic idea in string theory is quite simple. It says that the elementary constituents of matter are not point-like objects (particles) but one-dimensional objects. These one-dimensional objects, also known as the fundamental (or elementary) strings, have very specific properties which determine the various modes in which the string can vibrate. However, to the present day experimentalists these strings appear as particles since their size is small compared to the distance scale that can be probed by the

Fig. 6. Vibrating closed and open strings.

most powerful microscopes available today.[4] In particular, different vibrational states of a fundamental string appear to us as different elementary 'particles' just as the different modes of vibration of a single musical string can produce different harmonics of a note. Thus in string theory instead of having different types of elementary particles we have one single type of elementary string as the basic constituent of matter. Figure 6 shows some of the vibrational states of strings. As is evident from this figure, strings can come in two varieties — closed strings which have no boundary and open strings which have two end points forming its two boundaries.

Since quantum mechanics and special theory of relativity are two of the basic inputs in the standard model, and since string theory must include the standard model if it is to describe our universe, it is natural to require that string theory also respects the principles of quantum mechanics and special theory of relativity. However, one finds that for various technical reasons it is not easy to respect these principles. In fact, the only way we can respect these principles is by formulating string theory not in the usual three-dimensional space but in a hypothetical nine-dimensional space.[5] Furthermore in this nine-dimensional space one can formulate altogether five different types of string theory — known as the Type I, Type IIA, Type IIB, $E_8 \times E_8$ heterotic and SO(32) heterotic string theories. These five string theories differ from each other in the type of vibrations which the string performs. As a result they have different vibrational states, which is

[4]The most powerful microscopes available today are in fact the particle accelerators. In these machines we accelerate particles to a velocity close to that of light so that they carry very high energy and then collide them with other particles. This process has the capability of (indirectly) probing the structure of matter to a very small scale. The minimum distance that can be resolved by the current accelerators is about 10^{-16}cm.

[5]We often count time as an additional dimension and describe this as a ten-dimensional space-time. But in this article we shall only count the number of space dimensions.

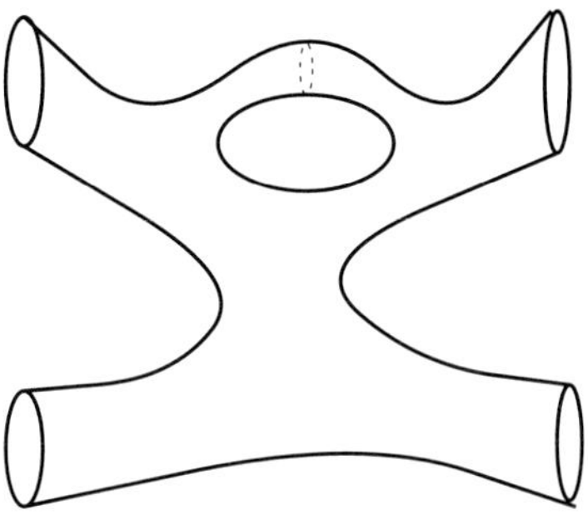

Fig. 7. A process describing a pair of strings scattering from each other.

reflected in the spectrum of elementary 'particles' that each of these theories produce.

Having nine space dimensions instead of three seems to be a serious problem. We shall return to this issue shortly and show that this, in fact, is not a very serious problem. However, let us leave aside this problem for a moment and discuss some of the good things which string theory provides. First of all, one finds that one of the vibrational states of string theory have properties identical to that of a graviton — the mediator of gravitational force. Furthermore one finds that string theory calculations do not suffer from any infinities of the type we encounter while trying to directly quantize general theory of relativity. Thus string theory provides us with a finite quantum theory of gravity!

It is instructive to try to understand why the probability amplitudes calculated in string theory are finite. For this we need to look at Fig. 7 describing the process of scattering of two strings. Like in the case of point particle theories, there are definite mathematical rules for calculating the probability amplitude of this process. The point to note is that in this diagram there are no points where specific events (like splitting of a single string into a pair of strings) take place; the diagram is completely smooth everywhere. As a result the divergences in the point particle theories — which arise when two or more such events take place at the same point at the same time — are absent in string theory. This is the intuitive reason why string amplitudes are finite.

At this point, we must mention that the graviton is only one of the many vibrational states of an elementary string. In fact, the laws of quantum mechanics tell us that a single elementary string has infinite number of vibrational states. Since each such vibrational state behaves as a particular

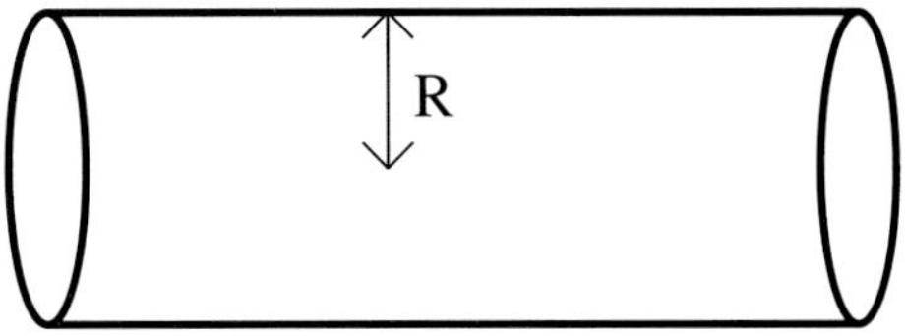

Fig. 8. A two-dimensional space with a compact coordinate.

type of elementary particle, string theory seems to contain infinite types of elementary particles. This would be in contradiction with what we observe in nature were it not for the fact that most of these elementary particles in string theory turn out to be very heavy, and not observable in present experiments. Thus, there is no immediate conflict between what string theory predicts and what we observe in actual experiments. On the other hand, these additional heavy elementary particles are absolutely essential for getting finite answers in string theory.

Let us now return to the issue about the dimension of space-time. Consistency of string theory demands that we can formulate the theory only in nine space dimensions. How can string theory be relevant for describing nature, which seems to have only three space dimension? The answer to this question is provided by an old idea known as compactification. This idea was pioneered by Kaluza and Klein during the first half of the twentieth century and Einstein himself had been attracted by this idea. We shall illustrate the basic idea by a simple example in which we begin with a world with two space dimensions instead of nine space dimensions. We take the two space coordinates to describe the surface of a cylinder of radius R instead of an infinite plane as shown in Fig. 8. All objects (including light) in this world can move only along the surface of the cylinder. Thus if we move along the vertical direction in the figure, then after traveling a certain distance ($2\pi R$ where R is the radius of the cylinder) we shall traverse the whole circumference of the circle and come back to the original point where we started. We call this a compact direction. In contrast, an object can travel along the horizontal direction without ever returning to its original position and we call this the non-compact direction.

Clearly if R is very large (larger than the range of the most powerful telescope) then the two-dimensional space will appear to be infinite in both directions and we would not know that one of the directions is compact. If R is within the visible range, then the two-dimensional creatures will start

Fig. 9. A two-dimensional space with a small compact coordinate.

seeing infinite number of images of each object separated by an interval of $2\pi R$ since light from any object can reach an observer in many (infinite number of) ways — directly, traveling once around the circumference, traveling twice around the circumference, etc. This may seem strange from our point of view but will not at all seem strange from the point of view of the two-dimensional people living in this world since they would always see their world this way. But now consider the case when R is very small, as shown in Fig. 9. Clearly this world looks one-dimensional as $R \to 0$. In fact, as long as R is smaller than the resolution of the most powerful microscope, the two-dimensional people will never know that they have a hidden dimension in their world. To them the world will appear to be one-dimensional.

This illustrates the way a universe with a certain number of space dimensions can 'appear to be' a universe with less number of dimensions. This idea can be generalized to make the nine-dimensional world of string theory look like the three-dimensional world in which we live. All we need to do is to take six of the nine space directions to be small, describing a compact space K. When the size of K is sufficiently small, the space will appear to be three-dimensional. The main difference with the two-dimensional example that we discussed is that while there is only one one-dimensional space (namely the circle) that can be used for making one direction compact, there are more possibilities in higher dimensions. An important class of six-dimensional spaces which are useful for compactification of string theory are the so-called Calabi–Yau spaces. There are many different six-dimensional Calabi–Yau spaces, and the theory that describes the three-dimensional world after compactification depends on the choice of the compact space K, as well as which of the five string theories we start from in nine dimensions.

Often the three-dimensional theory found this way comes very close to describing the world we see around us. In particular, when we examine the vibrational states of the string in such a space, not only do we find the graviton, but we often find 'gauge bosons' — the kind of particles which mediate strong, weak and electromagnetic forces. Some other vibrational states have properties similar to those of various quarks, leptons, Higgs particle, etc. Thus string theory has the potential of describing a unified theory of elementary particles and all the forces operating between them.

We would like to emphasize here that in string theory we use quantum mechanics and special theory of relativity as basic inputs; but the general theory of relativity and gauge theories come out of string theory. Thus string theory in a sense provides an explanation of why the forces operating in our universe are described by general theory of relativity and gauge theories.

Of course, all is not well at this stage. First of all, we have the problem that even though we know of many string compactifications which come very close to describing the world that we see, there is no known compactification that describes exactly the world that we see around us. Trying to look for a string compactification that describes exactly the theory that governs our universe is an active area of research in which many theorists are participating. Second, one might wonder what is the basic principle that one uses to decide which of the five string theories is the right theory for describing our universe. If we are looking for a theory that describes everything in our universe, wouldn't it be nicer to have a single mathematically consistent theory rather than five consistent theories? Finally, even if there is some principle that tells us which of the five string theories we should use, there are still many different choices of the compact space that bring us down to three dimension; and one might wonder what principle decides on the choice of the compact space. In fact, it is possible to have string compactification where the number of non-compact direction is different from three; all it requires to have d non-compact directions is to choose an appropriate compact space of dimension $(9 - d)$. Thus the question arises as to why our world is three-dimensional? We shall try to address some of these issues in the next section.

5. Duality, M-theory and the Early Universe

So far we have discussed the role played by the vibrational states of a single fundamental string. However, these are not the only possible objects in string theory. String theory contains many other types of objects which can be made of more than one (some time infinite number of) fundamental strings. We shall call these objects composite objects.

In conventional approach to the study of elementary constituents of matter, we make a clear distinction between elementary and composite objects. For example in the standard model the quarks are elementary particles while the proton and the neutron are composite particles made of quarks. The standard model tells us various properties of quarks and other elementary particles in the theory; the properties of protons, neutrons and other composite objects can be derived from the properties of these constituent particles. Thus elementary particles enjoy a previlaged position in the description of the theory.

The initial formulation of string theory was based on the same principle, with the role of elementary particles being taken over by the elementary strings. The vibrational states of the elementary string were the analogs of the elementary particles; all other objects made of more than one elementary string were composite objects whose properties could in principle be derived from the properties of the elementary string. However this picture, that gives a special role to the elementary particles, got modified dramatically after the discovery of duality symmetries in string theory. This is the story to which we now turn.

During the mid 90's it was realized that some time a pair of theories which 'look' different may actually describe the same physical theory. In other words, the same physical theory may have different descriptions as different compactifications of different string theories. This symmetry, relating the two apparently different theories, is known as the duality symmetry. This name is actually a misnomer, since often one finds more than two descriptions of the same physical theory. One of the surprising features of duality symmetries is that a particle which looks elementary in one description may appear as composite in a dual description. Thus whether a given particle is elementary or composite is not an intrinsic property of the particle, but depends on which particular description we use for the string theory under study.

Another aspect of duality is that typically the coupling constant of the theory — the parameter that determines the strength of various forces operating between the elementary particles — is related to the coupling constant of the dual theory in a complicated way. Due to this one finds that often a weakly coupled theory, i.e. a theory with small value of the coupling constant is related by duality to a theory with large value of the coupling constant. Since it is easier to do calculations in a theory for small value of the coupling constant, often duality relates the results of a complicated calculation in one theory to the results of a simple calculation in the dual theory.[6]

[6]Due to the difficulty in doing calculations in a strongly coupled theory, most of the dualities have not been proven, but have been tested in many different ways.

It is best to illustrate this with some examples. We begin with an example of duality involving theories with all nine dimensions non-compact. We had earlier introduced five different consistent string theories in nine dimensions. It turns out that the type I string theory and the $SO(32)$ heterotic string theory are dual to each other in the sense described above. They 'look' different because the set of elementary particles, obtained from the states of the elementary string, are quite different in the two theories. However when one considers the full set of particles — elementary and composite — in the two theories, one finds that the two sets are identical. The coupling constant of the heterotic string theory turns out to be equal to the inverse of the coupling constant of the type I theory. Thus when the heterotic string is weakly coupled the type I string is strongly coupled and vice versa.

Another example of duality involves string theories with five non-compact space directions. We take any one of the two heterotic string theories and take four of the space directions to be compact, each describing a circle of certain radius. Such a four-dimensional space is known as a four torus, denoted by the symbol T^4. On the other side we take type IIA string theory and make four of the space directions compact, this time describing a more complicated four-dimensional space known as K3. It turns out that these two five-dimensional string theories are dual to each other.

In special cases a particular compactification of string theory may be related to itself by a duality symmetry. In this case the duality symmetry will relate the elementary and composite particles in the same theory. Such theories are known called self-dual. For example type IIB string theory with all directions non-compact is a self-dual theory. Another example is any of the two heterotic string theories with six compact directions, each described by a circle. In both these theories duality typically relates an elementary particle to a composite particle.

Using various known dualities between different compactification of different string theories one can now argue that all five string theories are different ways of describing a single theory. This theory has been given the name M-theory. Different compactifications of different string theories which are not related by duality are to be regarded as different phases of M-theory, much in the same way that water, ice and steam are to be regarded as different phases of a single theory — the theory of water molecules.[7] A schematic (and much simplified) picture of the phases of

[7]One difference between these two cases is that while for water the three phases are stable for different values of temperature, pressure, etc., different compactifications of string theory are all stable phases at zero temperature.

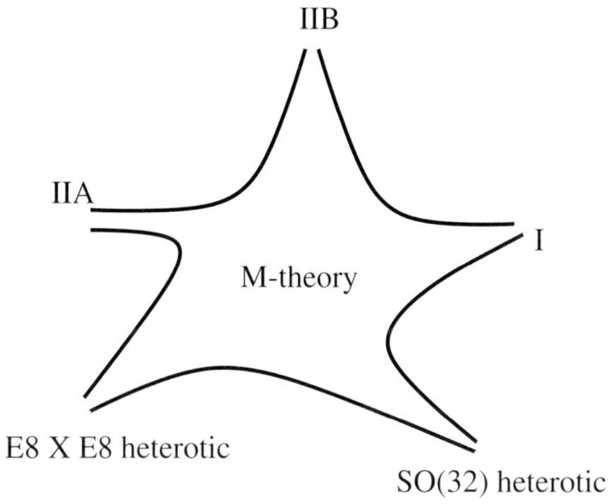

Fig. 10. Phases of M-theory.

M-theory has been shown in Fig 10. A point in this diagram represents a phase of M-theory, and the five holes represent the five weakly coupled string theories through which we may try to get a view of the different phases of the theory. In principle, any point can be viewed as an appropriate 'compactification' of any of the five string theories, but clearly if we consider a point near one of the windows — representing the corresponding string theory with small value of the coupling constant — we have a better view of the point from that window. Understanding what lies in the interior of the phase diagram representing phases of M-theory which cannot be viewed as weakly coupled theories from the viewpoint of any of the five string theories, is one of the most challenging problems for the present day string theorists.

Thus the problem of connecting M-theory to nature reduces to:

1. Demonstrating that there is a phase of M-theory that describes exactly the nature that we observe.
2. Explaining why nature exists in this particular phase and not in any other phase.

Both issues are currently under active investigation by many researchers. I shall end this talk by describing some speculative ideas on the second issue.

It has recently been found that M-theory has certain metastable phases. These metastable phases are analogous to the supercooled or superheated phases of matter. Consider for example the case of a supercooled water — water below the normal freezing point. As long as there is no disturbance the water remains as water, but a small disturbance in any part of the system will make a small region around that part condense into the more stable ice phase. This small region of ice will then expand inside the water and eventually convert the whole water into ice. Similarly the metastable phases of M-theory have the property that occasionally some regions of the universe in this phase may make transition into a more stable phase, and this region then grows with time, converting the surrounding region into the more stable phase.

There is however a crucial difference between the way a metastable phase of M-theory behaves and a metastable phase of a normal fluid behaves. The metastable phases of M-theory which are relevant for our discussion have an additional property that if any region of the universe is in that phase, it expands rapidly as a consequence of the laws of general theory of relativity. In technical terms we say that these phases have positive values of the cosmological constant — a constant that Einstein had introduced into the equations for general relativity and later abandoned due to lack of experimental evidence.[8] Often the rate of expansion of the universe due to this cosmological constant term turns out to be much faster than the rate of expansion of the bubbles of more stable phases which might form inside these metastable phases.

Let us now combine these two facts about the metastable phases of M-theory, and study how the universe will evolve if any region of the universe happens to be in such a metastable phase of M-theory. First of all, due to the cosmological constant term such a region of the universe will expand very rapidly. At the same time in different parts of the universe small regions of more stable phases will form[9] which will then grow, converting the surrounding region of the universe into the more stable phase. In fact, inside different bubbles we may have different stable phases of M-theory. In a normal fluid this process will stop when the walls of the expanding bubble eventually collide; and eventually the entire fluid will be converted to the most stable of all the phases. However, in the current situation this

[8]Recent experiments have found that our universe has a small but non-zero value of the cosmological constant. Thus Einstein was right after all! The phases of M-theory which we are discussing here have much larger values of the cosmological constant.

[9]Even if there is no external disturbance, the laws of quantum mechanics predict that there will be some intrinsic disturbance in the universe which causes some randomly chosen regions to form small bubbles of more stable phases.

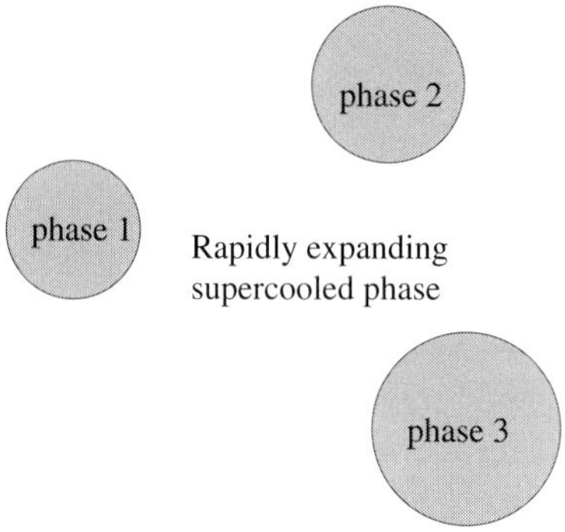

Fig. 11. State of the universe.

never happens since the universe as a whole is expanding rapidly due to the cosmological constant. Thus the process continues *ad infinitum*; the original universe keeps on expanding, and more and more bubbles of stable phases form in different regions of the universe. Eventually every possible phase of M-theory is realized inside one or more bubbles. This situation has been depicted in Fig. 11.

In this picture, no single phase of M-theory is preferred by nature. The world that we see around us exists in a particular phase simply because we happen to live in this part of the world. If we had lived in another part of the world we would see a different phase. Of course, in most of the phases of M-theory life as we know would be impossible, and so nobody would be there to observe these phases. But that is another matter!

6. Summary

There are various aspects of string theory which I have left out of our discussion. These include string theory analysis of black hole entropy, duality between string theory and gauge theory, etc. My main focus in this article has been to explain how string theory brings us closer to Einstein's dream. However, we are still quite far from realizing our final goal of finding a complete theory of elementary constituents of matter. It is up to the present and the future generation of string theorists to carry the theory forward towards this goal. It will be an uphill task but worth the effort.

References

[1] Zwiebach, B., *A First Course in String Theory* (Cambridge University Press, 2004).
[2] Johnson, C. V., *D-Branes* (Cambridge University Press, 2003).
[3] Polchinski, J., *String Theory,* Vols. 1 and 2 (Cambridge University Press, 1998).
[4] Green, M. B., Schwarz, J. H. and Witten, E., *Superstring Theory,* Vols. 1 and 2 (Cambridge University Press, 1987).

CHAPTER 4

Black Hole Entropy in String Theory: A Window into the Quantum Structure of Gravity

****

ATISH DABHOLKAR

Department of Theoretical Physics,
Tata Institute of Fundamental Research,
Homi Bhabha Road, Mumbai 400005, India

This article is intended as an impressionistic but reasonably self-contained account of black hole entropy, its physical significance, the tortuous historical route to its discovery, how it fits in the framework of string theory, and what we can learn from it about the fundamental degrees of freedom of quantum gravity.

1. Synopsis

One of the intriguing properties of a black hole is that it carries entropy much like an ordinary hot body. A beautiful general formula for this entropy due to Bekenstein and Hawking provides a deep connection between quantum mechanics, general relativity and thermodynamics.

For an ordinary body, its entropy equals the logarithm of the number of ways the atomic constituents of the body configure themselves. But for a black hole, it is far from clear what microscopic constituents might account for its entropy. It has been one of the outstanding open problems in physics to arrive at such a microscopic, statistical understanding of black hole entropy.

There has been considerable progress in recent years in addressing this question in the context of string theory. For a special class of black holes, in many cases, the number of microstates is **exactly** computable and is found to be in precise agreement with the number of states inferred from the entropy to **all orders** in a perturbative expansion. For this comparison to work, it is essential to systematically take into account quantum corrections to the spacetime geometry and the Bekenstein–Hawking formula itself.

Thus, the entropy of a black hole supplies us with precise quantitative information about the fundamental degrees of freedom and offers us glimpses of the inner workings of quantum gravity. These and related developments have led to important insights into the structure of quantum gravity which include in particular the notion of 'holography' and the emerging notion of 'quantum spacetime.'

Apart from its physical significance, the entropy of a black hole makes for a fascinating study in the history of science. It is one of the very rare examples where a scientific idea has gestated and evolved over several decades into an important conceptual and quantitative tool almost entirely on the strength of theoretical considerations. That we can proceed so far with any confidence at all with very little guidance from experiment is indicative of the robustness of the basic tenets of physics. It is therefore worthwhile to place black holes and their entropy in a broader context before coming to the more recent results pertaining to the quantum aspects of black holes within string theory.

This article is intended as an impressionistic but reasonably self-contained account of black hole entropy, its physical significance, the tortuous historical route to its discovery, how it fits in the framework of string theory, and what we can learn from it about the fundamental degrees of freedom of quantum gravity.

2. The Trinity of Constants

Perhaps a good measure of the unusual scope and influence of Einstein's ideas is the extent to which his thinking has shaped our understanding of the three fundamental constants of nature — the speed of light c, Planck's constant $\hbar$ and Newton's gravitational constant G. It is also revealing to see the extent to which these constants in turn have circumscribed the development of physics in the last century. In a sense, a very large part of modern physics can be viewed as an elucidation of the meaning of these constants and of the relation between them. With his Special and General Theory of Relativity and with his work in Quantum Theory, Einstein, more than any other single individual, has profoundly transformed the way we think about these constants.[1]

[1]It is equally remarkable that this does not exhaust the breadth of Einstein's oeuvre and leaves out his very important work in statistical physics including his work on Brownian motion and critical opalescence.

Of the three constants, Planck's constant $\hbar$, which governs the laws of the quantum world, has surely had a more pervasive influence on twentieth century physics. Even though Einstein never completely reconciled himself with the full implications of the quantum revolution that unfolded, his own contributions to the subject were nothing but revolutionary. In his paper in the miracle year 1905 on the photo-electric effect, Einstein introduced the light-quantum. With it, he introduced the corpuscular, quantum nature of the electromagnetic waves into physics and opened the door to the particle-wave duality of quantum mechanics. His other contributions to quantum theory were minor perhaps only by his monumental standards since they include the Bose–Einstein statistics and the idea of Bose–Einstein condensation of matter which was verified experimentally only very recently; his work on the specific heat of solids with which began the quantum theory of solids; his work on spontaneous and induced emission of radiation that anticipated quantum electrodynamics and led to the technology of lasers; and his critique of quantum mechanics with the Einstein–Podolsky–Rosen correlations which brought the spooky quantum behavior into sharp relief.

The second constant, the speed of light c is the cornerstone of the special theory of relativity. The fact that light is an electromagnetic wave traveling at the speed c was a celebrated piece of nineteenth century physics — a consequence of Maxwell's equations for the electromagnetic field. Lorentz and Poincaré, among others, had recognized that Maxwell's equations do not change their form under 'Lorentz transformations' which relate space and time coordinates of observers in uniform motion with respect to each other. This invariance of Maxwell's equations under Lorentz transformations however implied that time must dilate and lengths must contract. This was a startling conclusion. For Lorentz and Poincaré it signified a mysterious new dynamics. It sent them on a wrong track in a fruitless search for some complicated new forces that could explain the contraction of length.

Einstein's great insight was to look for the origin of Lorentz transformations not in the *dynamics*, which has to do with the forces, but in the *kinematics*, which has to do with the definition of time and length. He recognized that the dilation of time and contraction of space followed from a precise operational definition of 'simultaneity' of events that was purely kinematic. Since it was kinematic, it meant that the invariance under Lorentz transformations and the notion of space and time that it implied must be a property of the laws of motion of *all* objects and not only of the electromagnetic field. It is with this crucial observation of Einstein that the speed of light enters into particle mechanics. The relativistic kinematics puts the speed of light as the upper limit for *all* material propagation, particle or wave, electromagnetic or otherwise. The formula $E = mc^2$, immortalized in popular imagination, then follows from this new kinematics.

The third and the oldest of the trinity of constants, the gravitational constant G, belongs to the general theory of relativity, the greatest of Einstein's achievements. The general theory is actually not about G alone but rather about the two constants G and c together. Newton's law of gravitation requires that the force of gravity acts instantaneously. This is clearly at odds with special relativity which requires that no physical signal can travel faster than the speed of light. For example, according to Newton's gravity, if sun were to disappear suddenly, its gravity would disappear too and we on earth would come to know about it instantaneously, even though light takes about eight minutes to reach us from the sun. This state of affairs was clearly unsatisfactory and purely for reasons of internal consistency of the physical theory, it was essential to find a broader framework that synthesized Newton's gravity with special relativity. Such a framework would be required for describing phenomena where both G and c are important.

In general relativity, Newton's constant acquires a completely new meaning. For Newton, G is the constant of proportionality that appears in his inverse square law of gravitation. For Einstein, G is the constant that determines the degree to which a given distribution of matter warps space and time. In this new conception, spacetime was no longer a spectator of events but itself a dynamical participant that changed in response to the amount of matter present. It was no longer flat and Euclidean but curved in much the same way as the surface of the earth is round and curved. This curvature of spacetime is, according to Einstein, the origin of gravity. In a flat plane, parallel lines never meet. But in a curved space, as on the surface of the earth, two observers heading straight in two parallel lines starting on the equator will eventually meet at the north pole because of the curvature. In an analogous way, in general relativity, trajectories of two gravitating bodies appear to attract as if because of a force of gravity but it is only because they are moving in a curved spacetime.

The rich harvest of the synthesis effected by general relativity has still not been fully reaped. Just to take two examples of the major efforts in observational astronomy in this century, one is the LIGO project that is seeking to detect the wave of gravitational influence that travels at the speed of light and the other is the WMAP project that has already given us the incredibly detailed picture of the early big-bang cosmology within the framework of general relativity.

One unmistakable pattern in the history of modern physics is the progressive synthesis of ideas by which previously disparate structures are harmonized into a bigger framework. Thus, with special relativity, Einstein harmonized Maxwell's electrodynamics with Newton's mechanics introducing c into mechanics. With general relativity, he further harmonized this structure with Newton's law of gravitation, bringing together c and G.

Clearly one cannot stop here. The next synthesis requires harmonizing the special theory of relativity with quantum mechanics to describe the realm of phenomena where both c and $\hbar$ are important. It is the arena of relativistic quantum field theories that was developed over five decades in which Einstein himself did not play much of a role. Quantum field theory has proved to be the right framework where the duality of wave and particle nature of matter finds its full expression. Starting with Quantum Electrodynamics all the way to the Standard Model of Particle Physics based on quantum gauge theories, quantum field theory has occupied center stage in the study of fundamental interactions. Quantum field theory now encompasses three of the four fundamental interactions including electromagnetic as well as the weak and strong nuclear interactions. We possess a theory of elementary particles, the fundamental blocks of matter and their interactions, that has been tested to great accuracy to distances thousands of times smaller than the atomic nucleus.

That brings us to the final synthesis that still beacons us — a coherent description of physics in the realm where all three fundamental constants are simultaneously important. In other words, a quantum theory of gravity. Gravity, has thus far stubbornly refused to be integrated into the framework of quantum field theory. There is every indication that to do so, another revolutionary change in the paradigm of physics is necessary.

In this search for an overarching framework of quantum gravity that would harmonize quantum mechanics with general relativity, we have had little guidance from experiment. At this historical juncture, there is a peculiar situation in physics. We have two theories that are tremendously successful in their respective domain — Quantum field theory for describing the world at small scales in the realm of elementary particles, and General Relativity for describing the world at large scales all the way from our solar system to the universe. There is no experimental compulsion of an unexplained fact that forces us to bring these two theories together. At the same time, at a theoretical level it is absolutely necessary. As they stand, the two theories are in a violent conflict with each other in much the same way that special relativity was at odds with Newton's law of gravitation.

In the best of all possible worlds, theory and experiment work together. Without the sobering guidance from experiment the task of finding the correct theory of quantum gravity is much more difficult and far more risky. And yet history exhorts us to go on. Perhaps, Einstein's struggles towards the general theory of relativity can be our inspiration. Fortunately, we are also given an indirect but definitive piece of information that we can use to peer at the quantum structure of gravity.

It is the entropy of a black hole.

3. Black Holes

A black hole is a solution of Einstein's gravitational field equations in the absence of matter that describes the spacetime around a gravitationally collapsed star. Its gravitational pull is so strong that even light cannot escape it.

A black hole is now so much a part of our vocabulary that it can be difficult to appreciate the initial intellectual opposition to the idea of 'gravitational collapse' of a star and of a 'black hole' of nothingness in spacetime by several leading physicists, including Einstein himself.

To quote the relativist Werner Israel,

"*There is a curious parallel between the histories of black holes and continental drift. Evidence for both was already non-ignorable by 1916, but both ideas were stopped in their tracks for half a century by a resistance bordering on the irrational.*"

3.1. *Schwarzschild and Einstein*

On January 16, 1916, barely two months after Einstein had published the final form of his field equations for gravitation [1], he presented a paper to the Prussian Academy on behalf of Karl Schwarzschild [2], who was then fighting a war on the Russian front. Schwarzschild had found a spherically symmetric, static and exact solution of the full nonlinear equations of Einstein without any matter present.

The Schwarzschild solution was immediately accepted as the correct description within general relativity of the gravitational field outside a spherical mass. It would be the correct approximate description of the field around a star such as our sun. But something much more bizzare was implied by the solution. For an object of mass M, the solution appeared to become singular at a radius $R = 2GM/c^2$. For our sun, for example, this radius, now known as the Schwarzschild radius, would be about three kilometers. Now, as long as the physical radius of the sun is bigger than three kilometers, the 'Schwarzschild's singularity' is of no concern because inside the sun the Schwarzschild solution is not applicable as there is matter present. But what if the entire mass of the sun was concentrated in a sphere of radius smaller than three kilometers? One would then have to face up to this singularity.

Einstein's reaction to the 'Schwarzschild singularity' was to seek arguments that would make such a singularity inadmissible. Clearly, he believed, a physical theory could not tolerate such singularities. This drove him to write as late as 1939, in a published paper,

"*The essential result of this investigation is a clear understanding as to why the 'Schwarzschild singularities' do not exist in physical reality.*"

This conclusion was however based on an incorrect argument. Einstein was not alone in this rejection of the unpalatable idea of a total gravitational collapse of a physical system. In the same year, in an astronomy conference in Paris, Eddington, one of the leading astronomers of the time, rubbished the work of Chandrasekhar who had concluded from his study of white dwarfs, a work that was to earn him the Nobel prize later, that a large enough star could collapse.

It is interesting that Einstein's paper on the inadmissibility of the Schwarzschild singularity appeared only two months before Oppenheimer and Snyder published their definitive work on stellar collapse with an abstract that read,

"*When all thermonuclear sources of energy are exhausted, a sufficiently heavy star will collapse.*"

Once a sufficiently big star ran out of its nuclear fuel, then there was nothing to stop the inexorable inward pull of gravity. The possibility of stellar collapse meant that a star could be compressed in a region smaller than its Schwarzschild radius and the 'Schwarzschild singularity' could no longer be wished away as Einstein had desired. Indeed it was essential to understand what it means to understand the final state of the star.

3.2. *Event horizon*

What Einstein referred to as the 'Schwarzschild singularity' is in the matter of fact not a physical singularity at all. It is rather a coordinate singularity because of a bad choice of coordinates. The coordinates that Schwarzschild used to find his solution is more suited for an observer who wants to remain at a fixed distance r from the center. Far away, the constant r surface is time-like, that is, the observer who wants stay fixed at that radius is moving slowly compared to a freely falling observer. But near the Schwarzschild radius, $r = R$, because of the way the space time is curved in the Schwarzschild geometry, the surface $r = R$ is a light-like surface. That is, an observer who wants to remain fixed at that radius has to move at the speed of light. To do so, the observer has to turn on her rockets with infinite acceleration, a physical impossibility. It is this unphysical choice of coordinates that led to the misleading conclusion of a 'singularity' which is not really an intrinsic property of the geometry of spacetime.

Mathematically, a very close analogy for a such a coordinate singularity is the singularity in polar coordinates (ρ, θ) in a plane near the origin $\rho = 0$. The plane is perfectly flat at all points. Its origin is no different from any other point of the plane and the geometry of the plane at the origin is perfectly nonsingular. The proper coordinates at all points for a plane are the cartesian coordinates (x, y). These 'good' coordinates are related to the polar coordinates by $x = \rho \cos\theta$ and $y = \rho \sin\theta$. Now, at the origin, the polar coordinates are bad because the point $x = 0, y = 0$ does not have a unique coordinatization — as long as $\rho = 0$, all arbitrary values of θ would correspond to the same single point $x = 0, y = 0$. This coordinate singularity does not signify any intrinsic singularity of the geometry of the plane and in fact can be avoided by simply using the Cartesian coordinates near the origin.

The 'Schwarzschild singularity' can be similarly avoided by a proper choice of 'good' coordinates. In general relativistic spacetime, the analog of a cartesian coordinate frame is the coordinate frame of an observer who is freely falling through spacetime with her rocket engines switched off.

The surface $r = R$, even though not singular and perfectly ordinary in terms of its local geometry, is nevertheless rather peculiar in terms of the global causal structure. Since the surface is moving at the speed of light, once an observer crosses it, she cannot come out no matter how powerful her rockets. Because to do so, she would have to move faster than the speed of light. Thus, the $r = R$ surface is the boundary of the 'inside of a black hole' from behind which even light cannot escape to the observer who is sitting far away from the black hole. This boundary is then in the causal sense a one way surface. From outside, we can send signals across the surface but can never receive signals coming out from it. Such a one-way surface is called an 'Event Horizon'. The black hole is more precisely then the region of spacetime bounded by the event horizon. It is literally a hole in spacetime which is black because no light can come out of it. The name 'black hole' for this final state of the collapsed star, a spacetime with an event horizon, was proposed by John Wheeler in 1967 and it stuck.

Much of the interesting physics of a black hole, both classical and quantum, and the fact that a black hole has entropy, has to do with the existence of an event horizon.

3.3. *Simple and yet complex*

A black hole is at once the most simple and the most complex object.

It is the most simple in that it is completely specified by its mass, spin and charge. This remarkable fact is a consequence of the so-called 'No Hair Theorem'. For an astrophysical object like the earth, the gravitational field around it depends not only on its mass but also on how the mass is distributed and on the details of the oblateness of the earth and on the shapes of the valleys and mountains. Not so for a black hole. Once a star collapses to form a black hole, the gravitational field around it forgets all details about the star that disappears behind the even horizon except for its mass, spin and charge. In this respect, a black hole is very much like a structure-less elementary particle such as an electron.

And yet it is the most complex in that it possesses a huge entropy. In fact the entropy of a solar mass black hole is enormously bigger than the thermal entropy of the star that might have collapsed to form it. As we will see in Sec. 4, entropy gives an account of the number of microscopic states of a system. Hence, the entropy of a black hole signifies an incredibly complex microstructure. In this respect, a black hole is very unlike an elementary particle.

Understanding the simplicity of a black hole falls in the realm of classical gravity. By the early seventies, full fifty years after Schwarzschild, a reasonably complete understanding of gravitational collapse and of the properties of an event horizon was achieved within classical general relativity. The final formulation began with the singularity theorems of Penrose, area theorems of Hawking and culminated in the laws of black hole mechanics which we will come to in Sec. 5.

Understanding the complex microstructure of a black hole implied by its entropy falls in the realm of quantum gravity. To understand the meaning of the entropy of a black hole and its implications, let us first recall what we understand by entropy in thermodynamics and statistical physics.

4. Entropy and Microstates

Entropy is among the more subtle concepts in physics. It is not a property of a single microstate like energy or charge, but gives instead a count of the *total* number of microscopic states available to a macroscopic system that has fixed total energy and total charge.

Entropic and statistical considerations have been used to great advantage in physics to draw profound conclusions about the atomic microstructure from gross thermodynamic properties such as temperature and heat exchange. For example, already in the 19th century, some of the far-seeing physicists of the time were keenly aware of the crisis of classical physics based purely on statistical considerations.

For example, by looking at the specific heat of gases such as oxygen, Maxwell and Jeans had correctly concluded that classical molecular theory of gases was in serious trouble. The classical degrees of freedom of the theory implied much too large thermodynamic entropy. Similarly, Gibbs had inferred the strict quantum indistinguishability of oxygen molecules from considerations of thermodynamics and statistics. It is remarkable that these conclusions could be drawn at a time when full-fledged quantum mechanics was still several decades in the future. It is all the more remarkable that they were not based on subtle experiments as one might expect for a theory dealing with the atomic structure. Rather it was the logic of these enquiries which was subtle directed at explaining some gross thermodynamic feature of everyday gases such as their entropy and specific heat.

These are useful historical analogies to keep in mind as we look at the road ahead for quantum gravity. We dwell on these analogies a bit in this section to gain precise understanding of the relation between entropy and state-counting so that we can better appreciate the physical significance of the entropy of a black hole.

4.1. *Irreversibility and entropy*

Heat flows from a hot body to a cold body but not the other way around. How can we quantify this irreversibility of everyday experience?

The answer to this question came, among others, from a French engineer, Sadi Carnot who wanted to know how to build the most efficient steam engine to extract maximum possible work from it. He concluded that the most efficient engine is a reversible one, a thermodynamic analog of a frictionless engine. The efforts to quantify the notion of reversibility led to the notion of entropy. Define a quantity called entropy S as follows. If you add heat ΔQ to a body at temperature T, then the change in the entropy ΔS is given by

$$\Delta S = \frac{\Delta Q}{T} \,. \tag{1}$$

The important property of entropy is that in a reversible process, the total change in entropy is zero. Entropy is then an intrinsic property of a given system and is a function of the energy and the volume of the system. This allowed Carnot to enunciate the 'Second Law of Thermodynamics' which states that in an irreversible process entropy always increases.

The second law of thermodynamics explains the irreversibility of heat flow as follows. If heat $|\Delta Q|$ flows from a body at temperature T_1 to a body at temperature T_2, then the second body gains in entropy and the first body loses in entropy. The net change in entropy is then

$$\Delta S = |\Delta Q| \left(\frac{1}{T_2} - \frac{1}{T_1} \right) . \tag{2}$$

Since the second law requires $\Delta S > 0$ for an irreversible process, it implies that heat can flow only if $T_1 > T_2$.

At this stage, the second law is a phenomenological law. A microscopic understanding of the second law was completed by Boltzmann who gave a statistical interpretation of entropy.

4.2. *Entropy and disorder*

Boltzmann related the thermodynamic entropy S of a system to the total number Ω of different ways the microscopic constituents of the system can arrange themselves. He gave the fundamental relation

$$S = k \log \Omega \,, \tag{3}$$

where k is Boltzmann's constant.

Boltzmann's relation explains the second law of thermodynamics and the associated irreversibility from a microscopic point of view as a statistical tendency towards disorder. If you shake a jigsaw puzzle it is more likely to break than assemble itself simply because there are hugely more states when it is broken than when it is not. In other words, the system has more entropy when it is broken than when it is not. If we shake a puzzle in a box and look at it, statistically it is much more likely to be found in a broken state than in the assembled state. This explains why entropy always increases but only in a statistical sense.

To understand better the relation between thermodynamics and microscopic degrees of freedom, consider the entropy of oxygen in a room from this point of view. The thermodynamic entropy can be measured easily. To check Boltzmann's relation, we need to know Ω, the total number of ways for distributing N molecules of oxygen in a room of volume V. Now, a quantum particle like oxygen has wavelike nature and has a characteristic wavelength λ called its thermal de Broglie wavelength which can be thought of as its characteristic size. Since each molecule occupies volume λ^3, one can imagine that the room is divided into boxes, $\frac{V}{\lambda^3}$, in number. There are $\frac{V}{\lambda^3}$ ways a single molecule can be distributed in these boxes in the room. If we have N molecules, the total number would then be given by

$$\Omega = \frac{1}{N!}\left(\frac{V}{\lambda^3}\right)^N . \tag{4}$$

Here the crucial factor $N!$ is included because all oxygen molecules are identical. With λ the thermal de Broglie wavelength the logarithm of this quantity gives the correct answer for the thermodynamic entropy of a dilute gas like oxygen. This simple calculation was one of the great successes of molecular theory of gases in the nineteenth century which explained a gross, thermodynamic property in terms of a microscopic counting. It was in a sense a first peek at the atomic structure of matter.

4.3. *Quantum counting and classical overcounting*

There are a number of features of the formula (4) that are worth noting because they reveal important aspects of the concept of entropy and its physical significance. With Boltzmann's relation connecting this counting with the entropy, it already contains important hints about the quantum structure of matter.

First, without the factor $N!$, the Boltzmann relation would not be satisfied. This fact, first deduced by Gibbs from purely thermodynamic reasoning, attests to the strict indistinguishability of quantum oxygen molecules. Full theoretical understanding of this fact would require among other things, Bose–Einstein statistics and the spin-statistics theorem that came much later.

Second, in a strictly classical theory, in the limit $\hbar \to 0$, the thermal de Broglie wavelength would be zero. A classical point particle does not really occupy any space at all. As a result, the classical counting would give an infinity of states, and $k \log \Omega$ would be infinite even though physical entropy of a gas is finite. Finiteness of entropy is thus an indication of the quantum nature of the degrees of freedom.

This classical over-counting of degrees of freedom is typical and we will encounter it in the context of black holes as well. It manifests itself even in other statistical quantities. For example, the classical specific heat of oxygen is too large compared to the experimental value. This again is a consequence of a more subtle over-counting. Thinking of this problem, Maxwell remarked in a lecture given in 1875,

> "*Every additional degree of complexity which we attribute to the molecule can only increase the difficulty of reconciling the observed with the calculated value of the specific heat. I have now put before you what I consider the greatest difficulty yet encountered by the molecular theory.*"

Maxwell's difficulty had to do with the failure of classical equipartition theorem which assigns equal energy to all degrees of freedom. It was a thermodynamic manifestation of the inadequacy of classical ideas. Pondering over the same difficulty, Jeans made a prescient remark in 1890 that somehow 'the degrees of freedom seem to be frozen.'

In the full quantum theory which was to emerge several decades later, the resolution comes indeed from the fact at low temperature, average thermal energy would be much smaller than the quantum of energy needed to excite a degree of freedom such as a vibration of a molecule. In this case, such a degree of freedom is effectively frozen out as foreseen by Jeans. As a result, the classical equipartition theorem that Maxwell was using is not applicable thereby avoiding the conflict of theory with observation.

We are drawing this historical analogy to underscore the point that even when the full picture about the quantum theory of matter was very far from clear, it was possible to learn a great deal about the shape of the theory to come from this kind of thermodynamic and statistical considerations.

The situation with regards to quantum gravity is not quite the same but is in some ways analogous. Statistical reasoning has proved to be a valuable guide also in understanding the physics of black holes. One hopes that whatever may be the final form that the theory of quantum gravity takes, the insights that we can glean from the entropy of a black hole will be a part of it.

5. Black Hole Entropy

Before coming to the statistical aspects, let us first understand the thermodynamic aspects of a black hole.

5.1. *Bekenstein*

Jabob Bekenstein, then a graduate student of Wheeler, asked a simple-minded but incisive question [3]. What happens if you throw a bucket of hot water into a black hole? The entropy of the world outside the black hole would then decrease and the second law of thermodynamics would be violated. Should we give up this law that was won after half a century of hard struggle now in the presence of black holes?

Since the inside of the event horizon is never accessible by causal process to outside observers, whatever falls in it is forever lost. This fortunately does not affect the usual conservation laws of quantities such as energy and charge. For example, the energy of the bucket would be lost to the outside world but the energy or equivalently the mass of the black hole will go up by the same amount. The mass of the black hole can be measured from outside from its gravitational pull so if we keep track of the energy content of a black hole in our accounting of energy, then energy would continue to be conserved.

This suggested that even for entropy, if one could somehow associate an entropy with a black hole, then the second law of thermodynamics could be saved if we also keep track of the entropy of a black hole in our accounting of total entropy. But the 'No-Hair' theorem mentioned earlier showed that there were no other attributes of the black hole apart from its mass, charge and spin that could be measured from outside.

There is one quantity however, Bekenstein noted, namely the area of the black hole which behaved like entropy in many ways. For the Schwarzschild black hole, this is simply the area of the event horizon which equals $4\pi R^2$ where R is the Schwarzschild radius. For Bekenstein, the analogy was suggested by the remarkable laws of black hole mechanics, crystallized by Bardeen, Carter and Hawking, which had a striking resemblance with the three laws of thermodynamics for a body in thermal equilibrium.

Here A is the area of the horizon, M is the mass of the black hole, and κ is the surface gravity which can be thought of roughly as the acceleration at the horizon.[2]

[2]We have stated these laws for black holes without spin and charge but more general form is known.

Laws of Thermodynamics	**Laws of Black Hole Mechanics**
Temperature is constant throughout a body at equilibrium. T = constant.	Surface gravity is constant on the event horizon. κ =constant.
Energy is conserved. $dE = TdS$	Energy is conserved. $dM = \frac{\kappa}{8\pi} dA$
Entropy never decrease. $\Delta S \geq 0$	Area never decreases. $\Delta A \geq 0$

5.2. *Hawking radiation*

This analogy of Bekenstein was not immediately accepted because there was a serious difficulty with it. If a black hole has entropy and energy then it must also have temperature as can be seen from the definition of entropy (1). Now, any hot body must radiate and so also must a black hole with temperature. This conclusion was preposterous from the point of view of classical general relativity since after all a black hole was so named because it was perfectly black and nothing could come out of it.

Initially, Hawking among others, was willing to give up the second law in the face of this difficulty. Very soon though, he realized in his classic paper that a black hole could indeed have temperature once you include quantum effects [4]. In a quantum theory, virtual particles and antiparticles are constantly being created and annihilated from vacuum. Usually, they cannot be separated into real particles without violating conservation of energy because that would amount to creating a particle–antiparticle pair out of nothing. Near the event horizon, however, the antiparticle can fall into the black hole and the particle can escape to infinity as a real particle. Energy can be conserved in the process because the mass of black hole reduces accordingly. Hawking showed that the spectrum of these particles radiated from the black hole is exactly as if they are being radiated by a hot body at temperature T.

This temperature T of the black hole, now known as the Hawking temperature, is given by a simple formula

$$T = \frac{\hbar \kappa}{2\pi}, \tag{5}$$

where κ is the surface gravity encountered earlier. With this remarkable discovery, the table above becomes more than just an analogy. Indeed the left column is now precisely the same as the right column with the identification

$$S = \frac{Ac^3}{4\hbar G} = \frac{A}{4l^2}\,.$$

Here the length l is the Planck length 10^{-33} cm, a fundamental length constructed from the trinity of constants $l^2 = G\hbar/c^3$. This remarkably general formula is valid in all dimensions and for all kinds of black holes with mass, charge and spin.

Note that in the classical limit $\hbar \to 0$, the temperature vanishes, as it should since a black hole is really black classically. More importantly, in this limit entropy would become infinite. This is exactly as what we saw for oxygen gas in Sec. 4 and is the usual problem of classical overcounting. The finite quantum entropy of a black hole therefore signifies a certain discreteness of the degrees of freedom. This entropy is at present the only known physical quantity that involves all three fundamental constants of nature. It is therefore a precious clue about the microscopic structure of quantum gravity.

The discovery of thermodynamic entropy of a black hole in this way resolves the puzzle of Bekenstein about the apparent violation of the second law of thermodynamics. But it raises an even more interesting puzzle. Since the entropy of the black hole behaves in every respect like any other entropy that one encounters in statistical mechanics, what are the microstates of the black hole that can account for this thermodynamic entropy?

This remained an open problem for over two decades after Hawking's discovery. A complete understanding of the entropy of general black holes is still lacking, but there has been remarkable progress in addressing this question within the framework of string theory.

6. String Theory and Black Holes

Let us recall a few relevant facts about string theory[3] which is presently the leading candidate for a quantum theory of gravity.

String theory posits that the fundamental degrees of freedom are string-like extended objects instead of point-like elementary particles as assumed in quantum field theory. Different elementary particles arise as different oscillation modes of this fundamental string. Finding the spectrum of a

[3]For more details about string theory see the articles by David Gross and Ashoke Sen in this volume.

fundamental string is then analogous to finding what frequencies of sound will be produced by a sitar string. In this analogy, each musical note produced by the string would correspond to an elementary particle. One of the early surprises in the investigations of quantum string theories was that the spectrum of string theory always contained the graviton — the elementary particle that corresponds to a gravitational wave rippling through spacetime which carries the force of gravity. This striking fact was a natural consequence of the theory and was not put in by hand. Thus string theory is automatically a quantum theory of gravity. In a sense, quantum gravity is not only possible within string theory but is in fact necessary. Furthermore, when the gravitational coupling G is small, the interactions of the gravitons within string theory are free of the unphysical infinities that plagued earlier attempts to formulate quantum gravity within the framework of quantum field theory.

Earlier developments in string theory were limited to situations where gravitational interactions are weak. In the context of black holes however, the gravitational interactions are strong enough to warp spacetime into a black hole. Black holes therefore obtain a useful laboratory for testing the formalism of string theory beyond weak coupling. One of the striking successes of string theory is that for a special class of black holes, one can indeed explain the thermodynamic Bekenstein–Hawking entropy in terms of underlying microstates which can be counted exactly. What is more, in some examples, one can compute the corrections to the Bekenstein–Hawking entropy systematically to all orders in a perturbative expansion in inverse area and these too agree precisely with the microscopic counting.

The beautiful agreement that we find between the microscopic counting and the macroscopic, thermodynamic entropy not only resolves a long-standing puzzle raised about the interpretation of black hole entropy but also gives a strong hint that string theory provides a consistent framework for quantum gravity even at strong coupling.

6.1. *Counting black holes*

Supersymmetry is a generalization of Lorentz transformations. Just as special relativity requires that the laws of physics be invariant under Lorentz transformations, string theory requires that the laws of physics be invariant under a bigger symmetry, supersymmetry. Combining supersymmetry with Einstein's theory of gravity leads to a generalization of general relativity called supergravity.

The chief tool in dealing with the entropy of black holes in string theory is the spectrum of 'supersymmetric states.' Supersymmetric states of a theory are a special class of states that carry both mass and charge and have the property that their spectrum does not change as one changes the coupling constant of the theory. As a result, the number of such states can be counted reliably when Newton's constant G is small and gravitational interactions are weak. The counting is much easier in this limit as we will see below. Now, if one increases the value of G, then gravity becomes important, and a state with mass M and charges $Q_1, Q_2, \ldots$ undergoes gravitational collapse. Since, it is a supersymmetric state, the number of states at large G implied by the entropy of the corresponding black hole must equal the number of states counted at small G.

The most well-studied example that gives a microscopic account of the thermodynamic entropy is in five spacetime dimensions with three kinds of charges Q_1, Q_2, Q_3 [5]. In this case, it is possible to count the number of supersymmetric states with these charges and in the limit of large charges, number of such states $\Omega(Q_1, Q_2, Q_3)$ grows exponentially in a way that matches precisely with thermodynamic entropy $S(Q_1, Q_2, Q_3) \equiv \frac{A}{4l^2}$ of black holes with the same charges,

$$S(Q_1, Q_2, Q_3) = k \log \Omega(Q_1, Q_2, Q_3) = 2\pi k \sqrt{Q_1 Q_2 Q_3} \,.$$

6.2. *Black holes as strings*

To describe how this comparison is carried out, we consider instead a simpler system in four spacetime dimensions that has only two charges p and q. One advantage of this system is that the microscopic counting can be done more easily and exactly even for small charges. As a result, a much more detailed comparison can be carried out including all order corrections to the Bekenstein–Hawking formula.

String theory naturally lives in nine space dimensions. To obtain the physical space of three dimensions, it is necessary to 'compactify' or to curl up the extra six dimensions into small internal space. Thus, in string theory, one imagines that at each point in physical space there is attached a small ball of six dimensions. Now suppose that one of the directions of the internal space is a circle. Consider a string wrapping q times with momentum p along this circle. The string looks point-like in four dimensions. Usually along a string extending vertically, the oscillations can either move up or move down. In the type of string theory used in this context called the 'heterotic' string, if we have only up-moving oscillations with total energy $N = pq$, then this state is supersymmetric.

Now, a string extending along one of the nine spatial directions has eight transverse directions. In addition, in heterotic string theory, there are sixteen internal dimensions along which the string can carry up-moving oscillations. These extra internal dimensions have to do with the fact that in heterotic string, states can carry sixteen kinds of charges. Therefore, altogether, we have twenty-four up-moving oscillations. Each oscillation has frequency labeled by an integer, $n = 1, 2, 3, \ldots, \infty$ which basically counts the number of wavelengths of the oscillations that can fit on the circle traveling around the circumference. We need to distribute the total energy N among all these oscillators and find out how many ways there are of doing it to find the total number of states with charges p and q. This problem then maps to a well-known class of problems analyzed by Hardy and Ramanujan. The total number of our black hole states then equals the number of ways one can partition an integer N into a sum of integers, using integers of 24 different colors. This quantity is usually denoted as $p_{24}(N)$. For example, the total number of ways of partitioning the integer 5 using integers of only one color would be denoted $p_1(5)$. It is easy to find this number, since

$$5 = 1+1+1+1+1 = 2+1+1+1 = 2+2+1 = 3+1+1 = 3+2 = 4+1 = 5\,,$$

and hence $p_1(5) = 7$. It is also easy to see that this number grows very rapidly, in fact exponentially, as we increase either the integer or the number of colors at our disposal.

To count our black holes with large charges we then need to find $p_{24}(N)$ for large values of N. The answer can be found exactly to all orders,

$$\log \Omega(p,q) \sim 4\pi\sqrt{pq} - \frac{27}{2}\log\sqrt{pq} - \log\sqrt{2} - \frac{675}{32\pi\sqrt{pq}} - \frac{675\times 9}{2048\pi^2 pq} - \cdots . \tag{6}$$

How does this microscopic counting compare with the macroscopic entropy?

6.3. *Beyond Bekenstein and Hawking*

In this particular example considered above, it turns out, on the macroscopic side, classical spacetime is singular and the entropy vanishes. This seems to be in flat contradiction with the result from the microscopic counting. However, things get more interesting because string theory implies calculable corrections to general relativity. Einstein's equations are nonlinear partial differential equations that involve only two derivatives of the dynamical fields. From a modern perspective, these equations are expected to be just the low energy approximation and the full equations are expected to contain terms with higher derivatives of the dynamical fields coming from various quantum corrections. In the presence of the higher derivative terms, both the solution and the Bekenstein–Hawking formula itself gets modified. There exists an elegant generalization of the Bekenstein–Hawking formula due to Wald [6, 7] that correctly incorporates the effects of the higher derivative terms. It gives the entropy as an infinite series

$$S = a_0 A(Q) + a_1 \log A(Q) + \frac{a_2}{A(Q)} + \dots , \tag{7}$$

where the coefficients a_i can be computed explicitly from the specific form of the generalization of Einstein's equations that follows from string theory.

To apply Wald's formula one must first find the quantum corrections to Einstein's equations and then find the generalization of the Schwarzschild solution including these corrections. It would appear like an impossible task to solve these highly nonlinear, higher derivative partial differential equations. Fortunately, it turns out that using various available techniques and supersymmetry [8–11], it is possible to compute the higher derivative terms with precise numerical coefficients in string theory and find the corrected solution. One then finds that the solution including these quantum corrections has finite area [12, 13] and is given by $A = 8\pi l^2 \sqrt{pq}$ in terms of the charges p, q above and Planck length l. Using the Bekenstein–Wald–Entropy formula one then finds that the perturbative expansion in inverse area in (7) is the same as the expansion for large charges in $1/\sqrt{pq}$ in (6). The coefficients a_i can be computed exactly and are such that the entropy S in formula above match precisely with the infinite expansion of the microscopic counting (7) except for an additive constant that cannot yet be determined [12, 14].

7. Glimpses of Quantum Gravity

String theory is at present the only known framework for understanding black hole entropy in terms of counting albeit only in some special cases. The fact that even the corrections to the entropy can be understood in terms of microscopic counting to all orders is encouraging. It remains a challenge to see how these results can be extended to the Schwarzschild black hole without using the crutches of supersymmetry.

There are other important insights that have emerged from the study of black holes, most notably, the notion of 'Holography.' Bekenstein noted that the total number of degrees of freedom in a region must be proportional to the area of the region (and not its volume as one might naively expect) measured in units of the Planck length. Otherwise, black hole formation in this region would violate the second law of thermodynamics. This observation implies a dramatic reduction in the number of degrees of freedom of quantum gravity. It will take us too far afield to discuss these developments relating to holography in any detail here.

It is not clear yet what form the final formulation of quantum gravity will take but there is every indication that string theory will be a part it. In the absence of direct experimental evidence, one can subject the formalism of string theory to stringent tests of consistency. The striking agreement between thermodynamic, macroscopic properties of black holes and the microscopic structure of the theory assures us that we might be on the right track.

What would Einstein have thought of this road to quantum gravity? In his own research, he was a master of statistical reasoning and used it with incomparable skill to establish the quantum reality of atoms and light. He was also the one who gave us the theory of gravity based on the geometry of spacetime. Perhaps he would have appreciated the current struggles to learn about quantum gravity from the interplay between geometry and thermodynamics.

Acknowledgments

Some of the historical material in this article is drawn from *Subtle is the Lord*, by Abraham Pais and *Black Holes & Time Warps* by Kip Thorne where more detailed references can be found. For a review of black holes and related developments in string theory see the reviews [15–17].

References

[1] Einstein, A. [1915] *PAW*, p. 844.
[2] Schwarzschild, K. [1916] *PAW*, p. 189.
[3] Bekenstein, J. D. [1973] Black holes and entropy, *Phys. Rev.* **D7**, 2333–2346.
[4] Hawking, S. W. [1975] Particle creation by black holes, *Commun. Math. Phys.* **43**, 199–220.

[5] Strominger, A. and Vafa, C. [1996] Microscopic origin of the Bekenstein–Hawking entropy, *Phys. Lett.* **B379**, 99–104.
[6] Wald, R. M. [1993] Black hole entropy in the noether charge, *Phys. Rev.* **D48**, 3427–3431.
[7] Iyer, V. and Wald, R. M. [1994] Some properties of noether charge and a proposal for dynamical black hole entropy, *Phys. Rev.* **D50**, 846–864.
[8] Ferrara, S., Kallosh, R. and Strominger, A. [1995] $N = 2$ extremal black holes, *Phys. Rev.* **D52**, 5412–5416.
[9] Lopes Cardoso, G., de Wit, R. and Mohaupt, T. [1999] Corrections to macroscopic supersymmetric black-hole entropy, *Phys. Lett.* **B451**, 309–316.
[10] Lopes Cardoso, G., de Wit, B. and Mohaupt, T. [2000] Macroscopic entropy formulae and non-holomorphic corrections for supersymmetric black holes, *Nucl. Phys.* **B567**, 87–110.
[11] Ooguri, H., Strominger, A. and Vafa, C. [2004] Black hole attractors and the topological string.
[12] Dabholkar, A., Kallosh, R. and Maloney, A. [2004] A stringy cloak for a classical singularity, *JHEP* **12**, 059.
[13] Dabholkar, A. [2005] Exact counting of black hole microstates, *Phys. Rev. Lett.* **94**, 241301.
[14] Sen, A. [1995] Extremal black holes and elementary string states, *Mod. Phys. Lett.* **A10**, 2081–2094.
[15] David, J. R., Mandal, G. and Wadia, S. R. [2002] Microscopic formulation of black holes in string theory, *Phys. Rept.* **369**, 549–686.
[16] Mohaupt, T. [2000] Black holes in supergravity and string theory, *Class. Quant. Grav.* **17**, 3429–3482.
[17] Aharony, O., Gubser, S. S., Maldacena, J. M., Ooguri, H. and Oz, Y. [2000] Large n field theories, string theory and gravity, *Phys. Rept.* **323**, 183–386.

CHAPTER 5

The Winding Road to Quantum Gravity

****

ABHAY ASHTEKAR

Institute for Gravitational Physics and Geometry,
Physics Department, 104 Davey, Penn State,
University Park, PA 16802, USA
Max Planck Institut für Gravitationsphysik,
Albert Einstein Institut, 14476 Golm, Germany
Indian Academy of Sciences,
Sir C. V. Raman Road, Bangalore 560 060 India

The goal of this article is to present a brief history of quantum gravity for a general audience. While familiarity with basic ideas and notions of contemporary physics is assumed, technicalities are kept to a minimum and use of equations is avoided. Rather, the emphasis is on providing a coherent picture of the evolution of ideas and the current status of the subject.

1. The Beginning

General relativity and quantum theory are among the greatest intellectual achievements of the 20th century. Each of them has profoundly altered the conceptual fabric that underlies our understanding of the physical world. Furthermore, each has been successful in describing the physical phenomena in its own domain to an astonishing degree of accuracy. And yet, they offer us *strikingly* different pictures of physical reality. Indeed, at first, one is surprised that physics could keep progressing blissfully in the face of so deep a conflict. The reason is that phenomena for which *both* theories are essential occur at the Planck scale and the values of fundamental constants in our universe conspire to make the Planck length $\ell_{\rm Pl} = \sqrt{G\hbar/c^3} \sim 10^{-33}$cm truly minute and Planck energy $\mathrm{E}_{\rm Pl} = \sqrt{\hbar c/G} \sim 10^{19}$Gev absolutely enormous compared to laboratory scales. Thanks to this coincidence, we can happily maintain a schizophrenic attitude and use the precise, geometric picture of reality offered by general relativity while dealing with cosmological and astrophysical phenomena, and the quantum-mechanical world of chance and intrinsic uncertainties while dealing with atomic and subatomic particles. Clearly, this strategy is quite appropriate as a practical stand. But it is highly unsatisfactory from a conceptual viewpoint. Everything in our past experience in physics tells us that the two pictures we currently use must be approximations, special cases that arise as appropriate limits of a grander theory. That theory must therefore represent a synthesis of general relativity and quantum mechanics. This would be the quantum theory of gravity. The burden on this theory is huge: Not only should it correctly describe *all* the known gravitational processes, but it should also adequately handle the Planck regime. This is the theory that we invoke when faced with phenomena, such as the big bang and the final state of black holes, where the Planck scale is reached and worlds of general relativity and quantum mechanics unavoidably meet.

It may come as a surprise that the necessity of a quantum theory of gravity was pointed out by Einstein already in 1916 — barely a year after the discovery of general relativity. In a paper in the Preussische Akademie Sitzungsberichte, he wrote:

> *"Nevertheless, due to the inneratomic movement of electrons, atoms would have to radiate not only electromagnetic but also gravitational energy, if only in tiny amounts. As this is hardly true in Nature, it appears that quantum theory would have to modify not only Maxwellian electrodynamics but also the new theory of gravitation."*

Papers on the subject began to appear in the thirties most notably by Bronstein, Rosenfeld and Pauli. However, detailed work began only in the sixties. The general developments since then loosely represent four stages, each spanning roughly a decade.

First, there was the beginning: exploration. The goal was to do unto gravity as one would do unto any other physical field [8].[1] The electromagnetic field had been successfully quantized using two approaches: *canonical and covariant.* In the canonical approach, electric and magnetic fields obeying Heisenberg's uncertainty principle are at the forefront, and quantum

[1]Since this article is addressed to non-experts, except in the discussion of very recent developments, I will generally refer to books and review articles which summarize the state of the art at various stages of development of quantum gravity. References to original papers can be found in these reviews.

states naturally arise as (gauge-invariant) functions $\Psi(A)$ of the vector potential A on a constant time 3-surface of space-time. In the covariant approach on the other hand, one first isolates and then quantizes the two radiative modes of the Maxwell field in space-time, without carrying out a (3+1)-decomposition of space-time into space and time. The quantum states naturally arise as elements of the Fock space of photons. Attempts were made to extend these techniques to general relativity. In the electromagnetic case the two methods are completely equivalent. Only the emphasis changes in going from one to another. In the gravitational case, however, the difference is *profound*. This is not accidental. The reason is deeply rooted in one of the essential features of general relativity, namely the dual role of the space-time metric.

To appreciate this point, let us begin with field theories in Minkowski space-time, say Maxwell's theory to be specific. Here, the basic dynamical field is represented by a tensor field $F_{\mu\nu}$ on Minkowski space. The space-time geometry provides the kinematical arena on which the field propagates. The background, Minkowskian metric provides light cones and the notion of causality. We can foliate this space-time by a one-parameter family of constant-time three-planes, and analyze how the values of electric and magnetic fields on one of these surfaces determine those on any other surface. The isometries of the Minkowski metric let us construct physical quantities such as fluxes of energy, momentum, and angular momentum carried by electromagnetic waves. Geometry of Minkowski space, on the other hand, is fixed; it is completely insensitive to the properties of the electromagnetic field.

In general relativity, by contrast, there is no background geometry. The space-time metric itself is the fundamental dynamical variable. On the one hand, it is analogous to the Minkowski metric in Maxwell's theory; it determines space-time geometry, provides light cones, defines causality, and dictates the propagation of all physical fields (including itself). On the other hand, it is the analog of the Newtonian gravitational potential and therefore the basic dynamical entity of the theory, similar in this respect to the vector potential A_μ of the Maxwell theory. This dual role of the metric is in effect a precise statement of the equivalence principle that is at the heart of general relativity. It is this feature that is largely responsible for the powerful conceptual economy of general relativity, its elegance, its aesthetic beauty, its strangeness in proportion.

However, this feature also brings with it a host of problems. We see already in the classical theory several manifestations of these difficulties.

It is because there is no background geometry, for example, that it is so difficult to analyze singularities of the theory and to define the energy and momentum carried by gravitational waves. Since there is no *a priori* space-time, to introduce notions as basic as causality, time, and evolution, one must first solve the dynamical equations and *construct* a space-time. As an extreme example, consider black holes, whose definition requires the knowledge of the causal structure of the entire space-time. To find if the given initial conditions lead to the formation of a black hole, one must first obtain their maximal evolution and, using the causal structure determined by that solution, ask if its future infinity has a past boundary. If it does, space-time contains a black hole and the boundary is its event horizon. Thus, because there is no longer a clean separation between the kinematical arena and dynamics, in the classical theory substantial care and effort is needed even in the formulation of basic physical questions.

In quantum theory the problems become significantly more serious. To see this, recall first that, because of the uncertainty principle, already in non-relativistic quantum mechanics particles do not have well-defined trajectories; time-evolution only produces a probability amplitude, $\Psi(x,t)$, rather than a specific trajectory, $x(t)$. Similarly, in quantum gravity, even after evolving an initial state, one would not be left with a specific space-time. In the absence of a space-time geometry, how is one to introduce even habitual physical notions such as causality, time, scattering states, and black holes?

2. Early Developments

The canonical and the covariant approaches have adopted dramatically different attitudes to face these problems. In the canonical approach, one notices that, in spite of the conceptual difficulties mentioned above, the Hamiltonian formulation of general relativity is well-defined and attempts to use it as a stepping stone to quantization. The fundamental canonical commutation relations are to lead us to the basic uncertainty principle. The motion generated by the Hamiltonian is to be thought of as time evolution. The fact that certain operators on the fixed ('spatial') three-manifold commute is supposed to capture the appropriate notion of causality. The emphasis is on preserving the geometrical character of general relativity, on retaining the compelling fusion of gravity and geometry that Einstein created. In the first stage of the program, completed in the early sixties, the Hamiltonian formulation of the classical theory was worked out in detail by Dirac, Bergmann, Arnowitt, Deser and Misner and others [1, 4, 5, 12, 15]. The basic canonical variable was the 3-metric on a spatial slice. The ten Einstein's equations naturally decompose into two sets: four constraints on the metric and its conjugate momentum (analogous to the equation $\mathrm{Div}\vec{E} = 0$ of electrodynamics) and six evolution equations. Thus, in the Hamiltonian formulation, general relativity could be interpreted as the dynamical theory of 3-geometries. Wheeler therefore baptized it *geometrodynamics* [2, 3].

In the second stage, this framework was used as a point of departure for quantum theory. The basic equations of the quantum theory were written down and several important questions were addressed [3, 15]. Wheeler also launched an ambitious program in which the internal quantum numbers of elementary particles were to arise from non-trivial, microscopic topological configurations and particle physics was to be recast as 'chemistry of geometry'. However, most of the work in quantum geometrodynamics continued to remain formal; indeed, even today the field theoretic difficulties associated with the presence of an *infinite number of degrees of freedom* remain unresolved. Furthermore, even at the formal level, it has been difficult to solve the quantum Einstein's equations. Therefore, after an initial burst of activity, the quantum geometrodynamics program became stagnant. Interesting results were obtained in the limited context of quantum cosmology where one freezes all but a finite number of degrees of freedom. However, even in this special case, the initial singularity could not be resolved without additional 'external' inputs into the theory. Sociologically, the program faced another limitation: concepts and techniques which had been so successful in quantum electrodynamics appeared to play no role here. In particular, in quantum geometrodynamics, it is hard to see how gravitons are to emerge, how scattering matrices are to be computed, how Feynman diagrams are to dictate dynamics and virtual processes are to give radiative corrections. To use a well-known phrase [6], the emphasis on geometry in the canonical program "*drove a wedge between general relativity and the theory of elementary particles.*"

In the covariant[2] approach [5, 7, 9] the emphasis is just the opposite. Field-theoretic techniques are put at the forefront. The first step in this

[2]In the context of quantum gravity, the term 'covariant' is somewhat misleading because the introduction of a background metric violates diffeomorphism covariance. It is used mainly to emphasize that this approach does not involve a 3+1 decomposition of space-time.

program is to split the space-time metric $g_{\mu\nu}$ in two parts, $g_{\mu\nu} = \eta_{\mu\nu} + \sqrt{G}\, h_{\mu\nu}$, where $\eta_{\mu\nu}$ is to be a background, kinematical metric, often chosen to be flat, G is Newton's constant, and $h_{\mu\nu}$, the deviation of the physical metric from the chosen background, the dynamical field. The two roles of the metric tensor are now split. The overall attitude is that this sacrifice of the fusion of gravity and geometry is a moderate price to pay for ushering-in the powerful machinery of perturbative quantum field theory. Indeed, with this splitting most of the conceptual problems discussed above seem to melt away. Thus, in the transition to the quantum theory it is only $h_{\mu\nu}$ that is quantized. Quanta of this field propagate on the classical background space-time with metric $\eta_{\mu\nu}$. If the background is in fact chosen to be flat, one can use the Casimir operators of the Poincaré group and show that the quanta have spin two and rest mass zero. These are the gravitons. The Einstein–Hilbert Lagrangian tells us how they interact with one another. Thus, in this program, quantum general relativity was first reduced to a quantum field theory in Minkowski space. One could apply to it all the machinery of perturbation theory that had been so successful in particle physics. One now had a definite program to compute amplitudes for various scattering processes. Unruly gravity appeared to be tamed and forced to fit into the mold created to describe quantum electromagnetic interactions. Thus, the covariant quantization program was more in tune with the mainstream developments in physics at the time. In the early sixties, Gupta and Feynman outlined an extension of perturbative methods from quantum electrodynamics to gravity. A few years later DeWitt carried this analysis to completion by systematically formulating the Feynman rules for calculating scattering amplitudes among gravitons and between gravitons and matter quanta. He showed that the theory is unitary order by order in the perturbative expansion. By the early seventies, the covariant approach had led to several concrete results [7].

Consequently, the second stage of the covariant program began with great enthusiasm and hope. The motto was: Go forth, perturb, and expand. The enthusiasm was first generated by the discovery that Yang–Mills theory coupled to fermions is renormalizable (if the masses of gauge particles are generated by a spontaneous symmetry-breaking mechanism).[3] This led to a successful theory of electroweak interactions. Particle physics witnessed a renaissance of quantum field theory. The enthusiasm spilled over

[3] In fact DeWitt's quantum gravity work [7] played a seminal role in the initial stages of the extension of perturbative techniques from Abelian to non-Abelian gauge theories.

to gravity. Courageous calculations were performed to estimate radiative corrections. Unfortunately, however, this research soon ran into its first road block. The theory was shown to be non-renormalizable when two loop effects are taken into account for pure gravity and already at one loop for gravity coupled with matter [16]. To appreciate the significance of this result, let us return to the quantum theory of photons and electrons. This theory is perturbatively renormalizable. This means that, although individual terms in the perturbation expansion of a physical amplitude may diverge due to radiative corrections involving closed loops of virtual particles, these infinities are of a specific type; they can be systematically absorbed in the values of free parameters of the theory, the fine structure constant and the electron mass. Thus, by renormalizing these parameters, individual terms in the perturbation series can be systematically rendered finite. In quantum general relativity, such a systematic procedure is not available; infinities that arise due to radiative corrections are genuinely troublesome. Put differently, quantum theory acquires an infinite number of undetermined parameters. Although one can still use it as an effective theory in the low energy regime, regarded as a fundamental theory, it has no predictive power at all!

Buoyed, however, by the success of perturbative methods in electroweak interactions, the community was reluctant to give them up in the gravitational case. In the case of weak interactions, it was known for some time that the observed low energy phenomena could be explained using Fermi's simple four-point interaction. The problem was that this Fermi model led to a non-renormalizable theory. The correct, renormalizable model of Glashow, Weinberg and Salam agrees with Fermi's at low energies but marshals new processes at high energies which improve the ultraviolet behavior of the theory. It was therefore natural to hope that the situation would be similar in quantum gravity. General relativity, in this analogy, would be similar to Fermi's model. The fact that it is not renormalizable was taken to mean that it ignores important processes at high energies which are, however, unimportant at low energies, i.e. at large distances. Thus, the idea was that the correct theory of gravity would differ from general relativity but only at high energies, i.e. near the Planck regime. With this aim, higher derivative terms were added to the Einstein–Hilbert Lagrangian. If the relative coupling constants are chosen judiciously, the resulting theory does in fact have a better ultraviolet behavior. Stelle, Tomboulis and others showed that the theory is not only renormalizable but asymptotically free; it resembles the free theory in the high energy limit. Thus, the initial hope of 'curing' quantum general relativity was in fact realized. However, it turned out that the Hamiltonian of this theory is unbounded from below, and consequently the theory is drastically unstable! In particular, it violates unitarity; probability fails to be conserved. The success of the electroweak theory suggested a second line of attack. In the approaches discussed above, gravity was considered in isolation. The successful unification of electromagnetic and weak interactions suggested the possibility that a consistent theory would result only when gravity is coupled with suitably chosen matter. The most striking implementation of this viewpoint occurred in supergravity. Here, the hope was that the bosonic infinities of the gravitational field would be cancelled by those of suitably chosen fermionic sources, giving us a renormalizable quantum theory of gravity. Much effort went into the analysis of the possibility that the most sophisticated of these theories — $N = 8$ supergravity — can be employed as a genuine grand unified theory.[4] It turned out that some cancellation of infinities does occur and that supergravity is indeed renormalizable to two loops even though it contains matter fields coupled to gravity. Furthermore, its Hamiltonian is manifestly positive and the theory is unitary. However, it is believed that at fifth and higher loops it is again non-renormalizable.

3. Paradigm Shifts

By and large, the canonical approach was pursued by relativists and the covariant approach by particle physicists. In the mid-eighties, both approaches received unexpected boosts. These launched the third phase in the development of quantum gravity.

A group of particle physicists had been studying string theory to analyze *strong interactions* from a novel angle. The idea was to replace point particles by one-dimensional extended objects — strings — and associate particle-like states with various modes of excitations of the string. Initially there was an embarrassment: in addition to the spin-1 modes characteristic

[4]For a number of years, there was a great deal of confidence, especially among particle physicists, that supergravity was on the threshold of providing the complete quantum gravity theory. For instance, in the centennial celebration of Einstein's birthday at the Institute of Advanced Study, Princeton [13] — the proceedings of which were videotaped and archived for future historians and physicists — there were two talks on quantum gravity, both devoted to supergravity. A year later, in his Lucasian Chair inaugural address Hawking [14] suggested that end of theoretical physics was in sight because $N = 8$ supergravity was likely to be the final theory.

of gauge theories, string theory included also a spin-2, massless excitation. But it was soon realized that this was a blessing in disguise: the theory automatically incorporated a graviton. In this sense, gravity was already built into the theory! However, it was known that the theory had a potential quantum anomaly which threatened to make it inconsistent. In the mid-eighties, Green and Schwarz showed that there is an anomaly cancellation. Perturbative string theory could be consistent in certain space-time dimensions — 26 for a purely bosonic string and 10 for a superstring [19, 28]. Since strings were assumed to live in the background of Minkowski space-time, one could apply perturbative techniques. However, in this reincarnation, the covariant approach underwent a dramatic revision. Since it is a theory of extended objects rather than point particles, the quantum theory has brand new elements; *it is no longer a local quantum field theory.* The field theoretic Feynman diagrams are replaced by world-sheet diagrams. This replacement dramatically improves the ultraviolet behavior and, although explicit calculations have been carried out only at 2 or 3 loop order, it is widely believed that the perturbation theory is *finite* to all orders; it does not even have to be renormalized. The theory is also unitary. It has a single, new fundamental constant — the string tension — and, since various excited modes of the string represent different particles, there is a built-in principle for unification of all interactions![5] From the viewpoint of local quantum field theories that particle physicists have used in studying electroweak and strong interactions, this mathematical structure seems almost magical. Therefore there is a hope in the string community that this theory would encompass all of fundamental physics; it would be the 'theory of everything'.

Unfortunately, it soon became clear that string perturbation theory also faces some serious limitations. Perturbative finiteness would imply that each term in the perturbation series is ultra-violet finite.[6] However Gross and Periwal have shown that in the case of bosonic strings, when summed, the series diverges and does so uncontrollably. (Technically, it is not even 'Borel-summable'.) They also gave arguments that the conclusion would

[5]To date, none of the low energy reductions appears to correspond to the world we actually observe. Nonetheless, string theory has provided us with a glimpse of an entirely new vista: the concrete possibility that unification could be brought about by a tightly woven, non-local theory.

[6]But it does appear that there are infrared divergences. While this is an important limitation from the mathematical physics perspective, as in QED, these are regarded as 'harmless' for calculation of physical effects. I thank Ashoke Sen for discussions on this issue.

not be changed if one uses superstrings instead. Independent support for these arguments has come from work on random surfaces due to Ambjorn and others. One might wonder why the divergence of the sum should be regarded as a serious failure of the theory. After all, in quantum electrodynamics, the series is also believed to diverge. Recall however that quantum electrodynamics is an inherently incomplete theory. It ignores many processes that come into play at high energies or short distances. In particular, it completely ignores the microstructure of space-time and simply assumes that space-time can be approximated by a smooth continuum even below the Planck scale. Therefore, it can plead incompleteness and shift the burden of this infinity to a more complete theory. A 'theory of everything' on the other hand, has nowhere to hide. It cannot plead incompleteness and shift its burden. It must face the Planck regime squarely. So, if string theory is to be consistent, it must have key non-perturbative structures. The current and the fourth stage of the particle physics motivated approaches to quantum gravity is largely devoted to unravelling such structures and using them to address some of the outstanding physical problems.

On the relativity side, the third stage also began with an unexpected but innocuous-sounding observation: the geometrodynamics program laid out by Dirac, Bergmann, Wheeler and others simplifies significantly if we regard a connection — rather than the 3-metric — as the basic object [21]. While metrics determine distances and angles, connections enable one to 'parallel transport' objects along curves. A familiar example from the textbook quantum mechanics is the electromagnetic vector potential A that lets us transport the wave function $\Psi(x)$ of a charged particle, such as the electron, from one point to another along any given curve: under an infinitesimal displacement, while the change in the wave function of an uncharged particle is given just by $\Delta\vec{x} \cdot \vec{\nabla}\Psi(x)$, for a charged particle, it is given by $\Delta\vec{x} \cdot (\vec{\nabla} - (iq/\hbar)\,\vec{A})\Psi(x)$ where q is the charge of the particle. The presence of a non-zero $\vec{A}$ manifests itself in a change of phase of Ψ, the most dramatic example of which occurs in the celebrated Bohm–Aharanov effect. In QCD the (matrix-valued) vector potentials couple similarly to the wave functions of quarks and dictate the change of their state as one moves from one point to another. In the gravitational context, the most familiar connection is the one introduced by Levi–Civita which enables one to parallel transport a vector on a curved manifold. We now know that, in their quest for an unified field theory, Einstein and Schrödinger, among others, had recast general relativity as a theory of Levi–Civita connections (rather than metrics) already in the fifties [31]. However, the theory became rather complicated.

This episode had been forgotten and connections were reintroduced in the mid-eighties. However, now these were 'spin-connections', required to parallel propagate *spinors*, such as the left-handed fermions used in the standard model of particle physics [21, 24]. Rather than making the theory complicated, these connections *simplify* Einstein's equations considerably. For example, the dynamics of general relativity can now be visualized simply as a *geodesic motion* on the space of spin-connections (with respect to a natural metric extracted from the constraint equations). Since general relativity is now regarded as a dynamical theory of connections, this reincarnation of the canonical approach is called '*connection-dynamics*'.

Perhaps the most important advantage of the passage from metrics to connections is that the phase-space of general relativity is now the same as that of gauge theories [21, 24]. The 'wedge between general relativity and the theory of elementary particles' that Weinberg referred to is largely removed without sacrificing the geometrical essence of general relativity. One could now import into general relativity techniques that have been highly successful in the quantization of gauge theories. At the kinematic level, then, there is a unified framework to describe all four fundamental interactions. The dynamics, of course, depends on the interaction. In particular, while there is a background space-time geometry in electroweak and strong interactions, there is none in general relativity. Therefore, qualitatively new features arise. These were exploited in the late eighties and early nineties to solve simpler models — general relativity in 2+1 dimensions [21, 22]; linearized gravity clothed as a gauge theory [21]; and certain cosmological models. To explore the physical, 3+1 dimensional theory, a 'loop representation' was introduced by Rovelli and Smolin [26]. Here, quantum states are taken to be suitable functions of loops on the 3-manifold.[7] This led to a number of interesting and intriguing results, particularly by Gambini, Pullin and their collaborators, relating knot theory and quantum gravity [25]. Thus, there was rapid and unanticipated progress in a number of directions which rejuvenated the canonical quantization program. Since the canonical approach does not require the introduction of a background geometry or use of perturbation theory, and because one now has access to fresh, non-perturbative techniques from gauge theories, in relativity circles there is a hope that this approach may lead to well-defined, *non-perturbative* quantum general relativity (or its supersymmetric version, supergravity).

[7]This is the origin of the name 'loop quantum gravity'. The loop representation played an important role in the initial stages. Although this is no longer the case in the current, fourth phase, the name is still used to distinguish this approach from others.

However, a number of these considerations remained rather formal until the mid-nineties. Passage to the loop representation required an integration over the infinite dimensional space of connections and the formal methods were insensitive to possible infinities lurking in the procedure. Indeed, such integrals are notoriously difficult to perform in interacting field theories. To pay due respect to the general covariance of Einstein's theory, one needed diffeomorphism invariant measures and there were folk-theorems to the effect that such measures did not exist!

Fortunately, the folk-theorems turned out to be incorrect. To construct a well-defined theory capable of handling field theoretic issues, a *quantum theory of Riemannian geometry* was systematically constructed in the mid-nineties [36]. This launched the fourth (and the current) stage in the canonical approach. Just as differential geometry provides the basic mathematical framework to formulate modern gravitational theories in the classical domain, quantum geometry provides the necessary concepts and techniques in the quantum domain. Specifically, it enables one to perform integration on the space of connections for constructing Hilbert spaces of states and to define geometric operators corresponding, e.g. to areas of surfaces and volumes of regions (even though the classical expressions of these quantities involve non-polynomial functions of the Riemannian metric). There are no infinities. One finds that, at the Planck scale, geometry has a definite discrete structure. *Its fundamental excitations are one-dimensional, rather like polymers*, and the space-time continuum arises only as a coarse-grained approximation. The fact that the structure of space-time at Planck scale is qualitatively different from Minkowski background used in perturbative treatments reinforced the idea that quantum general relativity (or supergravity) may well be non-perturbatively finite.

Finally, quantum geometry is a general framework that is not tied down to general relativity (or supergravity). However, since general relativity is the best classical theory of gravity we have, it is well worth investigating, at least as the first step, whether quantum general relativity exists non-perturbatively. Much of research in loop quantum gravity has been focussed on this question. Quantum geometry effects have already been shown to resolve the big-bang singularity and solve some of the long-standing problems associated with black holes.

4. The Past Decade

The first three stages of developments in quantum gravity taught us many valuable lessons. Perhaps the most important among them is the realization that perturbative, field theoretic methods which have been so successful in other branches of physics are not as useful in quantum gravity. The assumption that space-time can be replaced by a smooth continuum at arbitrarily small scales leads to inconsistencies. We can neither ignore the microstructure of space-time nor presuppose its nature. We must let quantum gravity itself reveal this structure to us. Irrespective of whether one works with strings or supergravity or general relativity, one has to face the problem of quantization non-perturbatively. In the current, fourth stage both approaches have undergone a metamorphosis. The covariant approach has led to string theory and the canonical approach developed into loop quantum gravity. The mood seems to be markedly different. In both approaches, non-perturbative aspects are at the forefront and conceptual issues are again near center-stage. However, there are also key differences. Most work in string theory involves background fields and uses higher dimensions and supersymmetry as *essential* ingredients. The emphasis is on unification of gravity with other forces of Nature. Loop quantum gravity, on the other hand, is manifestly background independent. Supersymmetry and higher dimensions do not appear to be essential. However, it has not provided any principle for unifying interactions. In this sense, the two approaches are complementary rather than in competition. Each provides fresh ideas to address some of the key problems but neither is complete.

In the rest of this section, I will illustrate the current developments by sketching a few of the more recent results. In the case of string theory, my discussion will be very brief because these topics are discussed in greater detail by several other articles in this volume.

4.1. *String theory*

Over the past decade, novel non-perturbative ideas have been introduced in string theory. Unlike in the perturbative epoch, it is no longer a theory only of *one*-dimensional extended objects. Higher dimensional objects, called 'branes' have played an increasingly important role. Although for historical reasons it is still called 'string theory', from a fundamental, conceptual perspective, strings are no more basic than branes. Of particular interest are the *D-branes* introduced by Polchinski on which open strings satisfying

'Dirichlet type' boundary conditions can end (whence the adjective 'D'). These lie at the heart of the statistical mechanical calculation of entropy of large extremal black holes in string theory.

The second key development was even more radical: Maldecena made the bold proposal that string theory on a certain anti-De sitter background space-time is isomorphic with a gauge theory living on its boundary. In the first and the most studied version, the background space-time is assumed to be a product of a five-dimensional anti-De Sitter space-time with a five-dimensional sphere (whose radius equals the cosmological radius of the anti-De Sitter space-time) while the gauge theory lives on the four-dimensional boundary of the five dimensional anti-De Sitter space-time. Since then the setup has been generalized to various non-compact dimensions. The boundary conditions — and hence the resulting string theories — are not of direct physical interest because our universe has a positive, rather than a negative, cosmological constant and because the compact spheres do not represent microscopic 'curled-up' dimensions because they now have huge radii. Nonetheless, from a mathematical physics perspective, the proposed duality is fascinating because it relates a 'gravity theory' residing on a curved space-time with a qualitatively different 'gauge theory' living on a (conformally) flat space-time. It has had some powerful applications, e.g. in unravelling the structure of certain supersymmetric gauge theories through supergravity!

Finally there are interesting proposals relating various types of string theories that go under the name '*dualities*'. Although there are no conclusive proofs, these ideas suggest that the five perturbatively constructed string theories and supergravity may be special limits of a grander, unknown theory, generally referred to as the *M theory*. The scenario has generated a great deal of enthusiasm. For, the conjectured theory is likely to be very rich. In particular, it should provide isomorphisms between the strong coupling regime of one string theory to the weak coupling regime of another. Unravelling of its non-perturbative structures will undoubtedly provide qualitatively new insights and perhaps even *radically* change our current perspectives.

4.2. *Loop quantum gravity*

Over the past decade, the main thrust of research in loop quantum gravity has been on using quantum geometry to address some of the long standing problems in the field. Certain key techniques introduced by Thiemann have provided glimpses of the qualitative changes in quantum dynamics that occur because of the absence of a background geometry. Specifically, thanks to the fundamental discreteness of quantum Riemannian geometry, the ultraviolet divergences — also in the definition of matter Hamiltonians — are naturally tamed [36, 43]. However, in the full theory, two major issues still remain. First, there is a large number of ambiguities in the formulation of quantum Einstein's equations and one needs additional inputs to remove them. Second, it is still not clear whether any of the current formulations admits a semi-classical sector that reproduces the low energy world around us [36].[8] So far, advances of direct physical interest have occurred by adopting a strategy which has been effective also in string theory: isolate and analyze issues on which significant progress can be made in spite of the gaps in the understanding of the full theory. In the rest of this sub-section, I will illustrate how this strategy is implemented. Rather than describing several results briefly, I will focus just on one. This will enable me to provide some details that are necessary for the reader to appreciate the subtle manner in which quantum geometry effects operate.

The issue in question is the nature of the quantum big-bang. Most work in cosmology is carried out in the context of spatially homogeneous and isotropic models and perturbations thereof [41]. In the simplest model, the basic variables of the symmetry reduced classical system are the scale factor a and matter fields ϕ. Symmetries imply that space-time curvature goes as $\sim 1/a^n$, where $n > 0$ depends on the matter field under consideration. Einstein's equations then predict a big-bang, where the scale factor goes to zero and the curvature blows up. Space-time comes to an end and the classical physics stops. For over three decades a key question has been: Can these 'limitations' of general relativity be overcome in an appropriate quantum theory? In traditional quantum cosmologies, the answer is in the negative. Typically, to resolve the singularity one either has to use matter (or external clocks) with unphysical properties or introduce additional boundary conditions, e.g. by invoking new principles, that dictate how the universe began.

[8] Several different avenues are being pursued to address these issues [36]. These include the 'discrete approach' due to Gambini and Pullin [40] in which one discretizes the theory prior to quantization; spin-foam approaches due to Baez, Barrett, Crane, Perez, Rovelli and others [32, 37] in which one uses a background independent, the path integral analog of loop quantum gravity; and the 'master constraint program' of Dittrich and Thiemann [35] which uses some of the key ideas of Klauder's [33] affine quantum gravity program.

In a series of papers Bojowald, Ashtekar, Date, Hossain, Lewandowski, Maartens, Singh, Vandersloot and others have shown that the situation in loop quantum cosmology is quite different: the underlying quantum geometry makes a *qualitative* difference very near the big-bang [36, 38]. At first, this seems puzzling because after symmetry reduction, the system has only a *finite* number of degrees of freedom. Thus, quantum cosmology is analogous to quantum mechanics rather than quantum field theory. How then can one obtain qualitatively new predictions? The answer is quite surprising: if one follows the program laid out in the full theory, then even for the symmetry reduced model one is led to a new quantum mechanics! Specifically, the representation (of the observable algebra) that naturally arises in loop quantum cosmology is *inequivalent* to that used in the older, traditional quantum cosmology. And in the new representation, quantum evolution is well-defined right through the big-bang singularity.

More precisely, the situation in dynamics can be summarized as follows. Because of the underlying symmetries, dynamics is dictated just by one of the ten Einstein equations, called the Hamiltonian constraint. Let us consider the simplest case of homogeneous, isotropic cosmologies coupled to a scalar field. In traditional quantum cosmology, this constraint is the celebrated Wheeler–DeWitt equation [2, 3] — a second-order differential equation on wave functions $\Psi(a, \phi)$ that depend on the scale factor a and the scalar field ϕ. Unfortunately, some of the coefficients of this equations diverge at $a = 0$, making it impossible to obtain an unambiguous evolution across the singularity. In loop quantum cosmology, the scale factor naturally gets replaced by μ the momentum conjugate to the connection. μ ranges over the entire real line and is related to the scale factor via $|\mu| = \mathrm{const}\, a^2$. Negative values of μ correspond to the assignment of one type of spatial orientation, positive to the opposite orientation, and $\mu = 0$ corresponds to the degenerate situation at the singularity. The Wheeler–DeWit equation is now represented by a *difference equation* on the quantum state $\Psi(\mu, \phi)$:

$$C^+(\mu)\Psi(\mu + 4\mu_o,\, \phi) + C^o(\mu)\Psi(\mu\, \phi) + C^-(\mu)\Psi(\mu - 4\mu_o,\, \phi) = \ell_{\rm Pl}^2\, \hat{H}_\phi \Psi(\mu,\, \phi) \tag{1}$$

where $C^\pm(\mu), C^o(\mu)$ are fixed functions of μ; μ_o, a constant, determined by the lowest eigenvalue of the area operator and $\hat{H}_\phi$ is the matter Hamiltonian. Again, using the analog of the Thiemann regularization from the full theory, one can show that the matter Hamiltonian is a well-defined operator.

Primarily, (1) is the quantum Einstein's equation that selects the physically permissible $\Psi(\mu\,\phi)$. However, *if* we choose to interpret μ as a heuristic time variable, (1) can be interpreted as an 'evolution equation' which evolves the state through discrete time steps. The highly non-trivial result is that the coefficients $C^{\pm}(\mu), C^{o}(\mu)$ are such that *one can evolve right through the classical singularity*, i.e. right through $\mu = 0$. Since *all* solutions have this property, the classical singularity is resolved. However, to complete the quantization program, one has to introduce the appropriate scalar product on the space of solutions to the constraint, define physically interesting operators on the resulting Hilbert space $\mathcal{H}_{\rm final}$ and examine their expectation values and fluctuations, especially near the singularity.

All these steps have been carried out in detail in the case when ϕ is a massless scalar field[9] [47]. Specifically, in each classical solution, ϕ is a monotonic function of time. Therefore, one can regard it as an 'internal clock' with respect to which the scale factor evolves. With this interpretation, the discrete equation (1) takes the form $\partial_t^2\Psi = -\Theta\Psi$, where Θ is a self-adjoint operator, independent of $\phi \sim t$. This is precisely the form of the Klein–Gordon equation in static space-times. (In technical terms, this provides a satisfactory 'deparametrization' of the theory.) Therefore, one can use techniques from quantum field theory in static space-times to construct an appropriate inner product and define a complete family of ('Dirac') observables. Using the two, one can construct semi-classical states — analogs of coherent states of a harmonic oscillator — and write down explicit expressions for expectation values and fluctuations of physical observables in them. As one might expect, the evolution is well-defined across the singularity but quantum fluctuations are huge in its neighborhood.

Now that there is a well-defined theory, one can use numerical methods to evolve quantum states and compare quantum dynamics with the classical one in detail. Since we do not want to make *a priori* assumptions about what the quantum state was at the big-bang, it is best to start the evolution not from the big bang but from late times ('now'). Consider then wave functions which are sharply peaked at a classical trajectory at late times and evolve them backward. The first question is: how long does the state remain semi-classical? A pleasant surprise is that it does so till *very* early times — essentially till the epoch when the matter density reaches the

[9]The extension of the analysis to include potential terms for the matter field ϕ or anisotropies for the combined system involves only technical complications. The overall conceptual picture remains the same.

Planck density. Now, this is precisely what one would physically expect. However, with a complicated difference equation such as (1), *a priori* there is no guarantee that semi-classicality would not be lost very quickly. In particular, this result provides support for the standard practice, e.g. in inflationary models, of assuming a classical continuum in the very early universe. Next, one can ask what happens to the quantum state very near and beyond the big-bang. As explained above, the state loses semi-classicality (i.e. fluctuations become large) near the big-bang. Does it then remain in a 'purely quantum regime' forever or does it again become semi-classical beyond a Planck regime on the 'other side' of the big bang? This is a question that lies entirely outside the domain of the standard Wheeler–Dewitt equation because it loses predictivity at the big-bang. In loop quantum cosmology, on the other hand, the evolution is well-defined and completely deterministic also beyond the big-bang. *A priori* there is no way to know what the answer would be. Space-time may well have been a 'quantum foam' till the big-bang and classicality may then have emerged only after the big-bang. Or, there may have been a classical space-time also on the 'other side'. Detailed numerical calculations show that the wave function becomes semi-classical again on the other side; *gravity becomes repulsive in the Planck regime, giving rise to a 'bounce'.* Thus, loop quantum cosmology predicts that the universe did not originate at the big bang but has a long prior history. Through quantum dynamics, the universe tunnels from a contracting phase in the distant past ('before the bang') to an expanding phase in the distant future ('now') in a specific manner. Classically, of course such a transition is impossible.

To summarize, the infinities predicted by the classical theory at the big-bang are artifacts of assuming that the classical, continuum space-time approximation is valid right up to the big-bang. In the quantum theory, the state can be evolved through the big-bang without any difficulty. However, the classical, continuum completely fails near the big-bang; figuratively, the classical space-time 'dissolves'. This resolution of the singularity without any 'external' input (such as matter violating energy conditions) is dramatically different from what happens with the standard Wheeler–DeWitt equation of quantum geometrodynamics [2–5, 8]. However, for large values of the scale factor, the two evolutions are close; as one would have hoped, quantum geometry effects intervene only in the 'deep Planck regime' resulting in a quantum bridge connecting two classically disconnected space-times. From this perspective, then, one is led to say that the most striking of the consequences of loop quantum gravity are not seen in

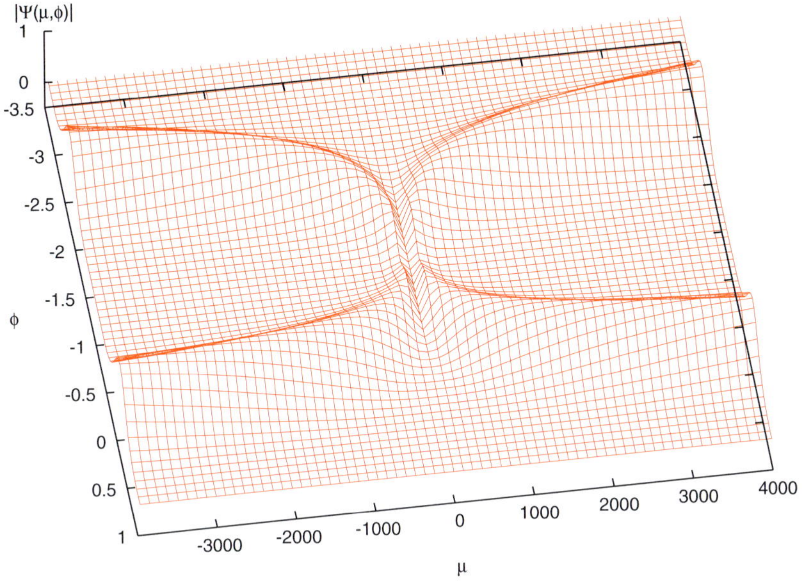

Fig. 1. Comparison between quantum and classical evolutions via plot of $|\Psi(\mu, \phi)|$. Since $\mu \to -\mu$ changes only the spatial orientation, it suffices to consider just $\mu \geq 0$. Except in the Planck regime very near $\mu = 0$, Ψ is sharply peaked at the classical trajectories. But the trajectory in the top half represents an *expanding universe* while that in the bottom half, a *contracting universe*. Thus, quantum geometry in the Planck regime bridges two vast but classically disjoint space-times.

older approaches because they 'wash out' the fundamental discreteness of quantum geometry.

5. Outlook

The road to quantum gravity has been long, spanning some four decades. Along the way came many new insights, jubilations as well as frustrations. Because of the page limit, I could only provide a general flavor of these trials, tribulations and triumphs. In particular, I had to restrict myself to the 'main-stream' programs whose development can be continuously tracked over several decades. There also exist a number of other fascinating and highly original approaches — particularly causal dynamical triangulations [27, 39], Euclidean quantum gravity [11, 42], twistor theory [10, 20] and the related theory of H-spaces [17], asymptotic quantization [18], non-commutative geometry [23] and causal sets [34] — that I could not discuss.[10]

But I hope I have managed to convey that, in spite of all the twists and turns in the winding road, there have been definite advances. We have learned that, because relativistic gravity is so deeply intertwined with space-time geometry, quantum gravity has unforeseen dimensions that would have surprised even the great leaders of the early period. We tried hard to extend perturbative methods of local quantum field theory which have been so successful in QED [8]. The efforts did lead to a successful perturbative framework for non-Abelian gauge theories [7, 16]. But we found conclusive evidence that these methods are insufficient for quantum gravity: they lead to uncontrollable ultraviolet divergences [16]. Thanks to string theory, we now have, for the first time, a concrete alternative — a computational framework to calculate scattering amplitudes which yields finite results to any order in perturbation theory [28]. Furthermore, the theory provides a brand new avenue to the unification of all interactions; the plethora of elementary particles is now reduced to various vibrational modes of the superstring. For decades we have been troubled by the fact that space-time of general relativity comes to an abrupt end at singularities and classical physics literally stops there. Loop quantum gravity has shown that this is an artifact of pushing the classical theory beyond the domain of its validity [36, 38]. Quantum geometry extends its life. What we thought of as a 'tiny, Planck scale region' can actually be a bridge joining our space-time to another vast classical region [47]; quantum space-time may be vastly larger than what general relativity had us believe.[11] Finally, both approaches have provided fascinating insights into the nature of quantum black holes [29, 36], a topic that would require a separate article in its own right.

All currently active directions point to the necessity of radical revisions of the 20th century paradigm of theoretical physics. String theory abandons local quantum field theories altogether and focuses instead on the study of extended quantum objects. Loop quantum gravity asks us to forego our cherished space-time continuum and embrace a quantum geometry instead. Twistor theory and non-commutative geometry suggest that we abandon the familiar space-time already at the classical level and reformulate general relativity well before the word 'quantum' is uttered or the symbol $\hbar$ introduced. No matter which of these approaches find an expression in the

[10] Accounts of the present status of several of these approaches can be found in the articles by Dowker, Ford, Gambini and Pullin and Penrose in [46].

[11] In Sec. 4.2 I discussed cosmological singularities. However, the situation is similar also for space-like black hole singularities [44, 45].

final quantum gravity theory, it is clear that quantum gravity will have deep ramifications on all of fundamental physics. General relativity led to a radical revision of our notions of space and time, thereby reshaping the conceptual foundation of *all of physics*. Impact of the successful quantum gravity theory on fundamental physics will be even deeper.

But today we cannot be certain which — or indeed, any — of these directions will constitute major components of the final theory. Thanks to the sustained work spanning many decades, most notable advances have occurred in the 'covariant' and 'canonical' approaches. However, even here one encounters serious incompleteness and some troubling features. How is the supersymmery broken in string theory and how does reduction to four large dimensions occur? While the theory is very tight in terms of its fundamental constants, there is a *huge* freedom in the choice of 'moduli-parameters'. There appear to be over 10^{200} 'vacua', each giving rise to its own low energy theory! How is this freedom to be reduced? No compelling principle seems to be in sight. More generally, our understanding of the presumed M-theory is very incomplete. Incompleteness pervades also loop quantum gravity. How are the ambiguities in the formulation of quantum Einstein's equations removed in the full theory? Does this theory admit a viable semi-classical sector? Through minisuperspaces we have learned that cosmological and black hole singularities are resolved through quantum geometry effects and loop quantum gravity enables one to perform a deterministic evolution across these singularities. Do these features survive beyond the minisuperspace approximation? How do inhomogeneous perturbations evolve in the cosmological context? Is this evolution compatible with observational constraints on structure formation? While there is vigorous ongoing research to answer such questions in both approaches, one cannot say that a satisfactory resolution is imminent. Even more important is the issue of observations. So far, not a single non-trivial and firm prediction of any quantum gravity theory has been verified directly. Therefore, as we celebrate the 100th anniversary of Einstein's Annus Mirabilis it is important that we maintain a long range perspective and not repeat our past error of overconfidence (see footnote 4). In particular, we would do well to avoid the traps that the celebrated biologist François Jacob [30] warned all scientists about:

> *The danger for scientists is not to measure the limits of their science, and thus their knowledge. This leads to mix what they believe and what they know. Above all, it creates the certitude of being right [prematurely].*

Acknowledgments

My understanding of quantum gravity has deepened through discussions with a large number of colleagues. Among them, I would especially like to thank John Baez, Peter Bergmann, Martin Bojowald, Alex Corichi, Klaus Fredenhagen, Rodolfo Gambini, Jim Hartle, Gary Horowitz, Ted Jacobson, Kirill Krasnov, Jerzy Lewandowski, Don Marolf, Jose Mourão, Hermann Nicolai, Roger Penrose, Jorge Pullin, Carlo Rovelli, Hanno Sahlmann, Ashoke Sen, Lee Smolin, Thomas Thiemann, and Madhavan Varadarajan. I would like to thank Naresh Dadhich, Shyam Date and Gary Horowitz for comments that improved the historical account. This work was supported in part by the NSF grant PHY 0090091, the Alexander von Humboldt Foundation, the Eberly research funds of The Pennsylvania State University and the Sir C.V. Raman Chair of the Indian Academy of Sciences.

References

[1] Arnowitt, R., Deser, S. and Misner, C. W. [1962] The dynamics of general relativity, in *Gravitation: An Introduction to Current Research*, ed. Witten L. (John Wiley, New York).
[2] Wheeler, J. A. [1962] *Geometrodynamics* (Academic Press, New York).
[3] Wheeler, J. A. [1964] Geometrodynamics and the issue of the final state, *Relativity, Groups and Topology*, eds. DeWitt, C. M. and DeWitt, B. S. (Gordon and Breach, New York).
[4] Komar, A. [1970] Quantization program for general relativity, in *Relativity*, eds. Carmeli, M., Fickler, S. I. and Witten, L. (Plenum, New York).
[5] Ashtekar, A. and Geroch, R. [1974] Quantum theory of gravitation, *Rep. Prog. Phys.* **37**, 1211–1256.
[6] Weinberg, S. [1972] *Gravitation and Cosmology* (John Wiley, New York).
[7] DeWitt, B. S. [1972] Covariant quantum geometrodynamics, in *Magic Without Magic: John Archibald Wheeler*, ed. Klauder, J. R. (W. H. Freeman, San Fransisco).
[8] Isham, C. J. [1975] An introduction to quantum gravity, in *Quantum Gravity, An Oxford Symposium*, eds. Isham, C. J., Penrose, R. and Sciama, D. W. (Clarendon Press, Oxford).
[9] Duff, M. [1975] Covariant quantization in *Quantum Gravity, An Oxford Symposium*, eds. Isham, C. J., Penrose, R. and Sciama, D. W. (Clarendon Press, Oxford).
[10] Penrose, R. [1975] Twistor theory, its aims and achievements, *Quantum Gravity, An Oxford Symposium*, eds. Isham, C. J., Penrose, R. and Sciama, D. W. (Clarendon Press, Oxford).

[11] Hawking, S. W. and Israel, W. (eds.) [1980] *General Relativity, An Einstein Centenary Survey* (Cambridge University Press, Cambridge).
[12] Bergmann, P. G. and Komar, A. [1980] The phase space formulation of general relativity and approaches toward its canonical quantization, *General Relativity and Gravitation Vol. 1, On Hundred Years after the Birth of Albert Einstein*, ed. Held, A. (Plenum, New York).
[13] Wolf, H. (ed.) [1980] *Some Strangeness in Proportion* (Addison Wesley, Reading).
[14] Hawking, S. W. [1980]. *Is End in Sight for Theoretical Physics?: An Inaugural Address* (Cambridge University Press, Cambridge).
[15] Kuchař, K. [1981] Canonical methods of quantization, in *Quantum Gravity 2, A Second Oxford Symposium*, eds. Isham, C. J., Penrose, R. and Sciama, D. W. (Clarendon Press, Oxford).
[16] Isham, C. J. [1981] Quantum gravity — An overview, in *Quantum Gravity 2, A Second Oxford Symposium*, eds. Isham, C. J., Penrose, R. and Sciama, D. W. (Clarendon Press, Oxford).
[17] Ko, M., Ludvigsen, M., Newman, E. T. and Tod, P. [1981] The theory of $\mathcal{H}$ space, *Phys. Rep.* **71**, 51–139.
[18] Ashtekar, A. [1984] *Asymptotic Quantization* (Bibliopolis, Naples); also available at http://cgpg.gravity.psu.edu/research/asymquant-book.pdf.
[19] Greene, M. B., Schwarz, J. H. and Witten, E. [1987] *Superstring Theory*, Vols. 1 and 2 (Cambridge University Press, Cambridge).
[20] Penrose, R. and Rindler, W. [1988] *Spinors and Space-Times*, Vol. 2 (Cambridge University Press, Cambridge).
[21] Ashtekar, A. [1991] *Lectures on Non-Perturbative Canonical Gravity*, Notes prepared in collaboration with R. S. Tate (World Scientific, Singapore).
[22] Ashtekar, A., Mathematical problems of non-perturbative quantum general relativity in *Gravitation and Quantizations: Proceedings of the 1992 Les Houches Summer School*, eds. Julia, B. and Zinn-Justin, J. (Elsevier, Amsterdam); also available as `gr-qc/9302024`.
[23] Connes, A. [1994] *Non-Commutative Geometry* (Academic Press, New York).
[24] Baez, J. and Muniain, J. P. [1994] *Gauge Fields, Knots and Gravity* (World Scientific, Singapore).
[25] Gambini, R. and Pullin, J. [1996] *Loops, Knots, Gauge Theories and Quantum Gravity* (Cambridge University Press, Cambridge).
[26] Rovelli, C. [1998] Loop quantum gravity, *Living Rev. Rel.* **1**, 1.
[27] Loll, R. [1998] Discrete approaches to quantum gravity in four dimensions, *Living Rev. Rel.* **1**, 13.
[28] Polchinski, J. [1998] *String Theory*, Vols. 1 and 2 (Cambridge University Press, Cambridge).
[29] Horowitz, G. T. [1998] Quantum states of black holes, in *Black Holes and Relativistic Stars*, ed. Wald, R. M. (University of Chicago Press, Chicago).
[30] Jacob, F. [1998] *Of Flies, Men and Mice* (Harvard University Press, Boston).

[31] Ashtekar, A. [2000] Quantum mechanics of geometry, in *The Universe: Visions and Perspectives*, eds. Dadhich, N. and Kembhavi, A. (Kluwer Academic, Dordretch); gr-qc/9901023.

[32] Perez, A. [2003] Spin foam models for quantum gravity, *Class. Quant. Grav.* **20**, R43–R104.

[33] Klauder, J. [2003] Affine quantum gravity, *Int. J. Mod. Phys.* **D12**, 1769–1774.

[34] Sorkin, R. [2003] Causal sets: Discrete gravity, gr-qc/0309009.

[35] Thiemann, T. [2003] The phoenix project: Master constraint programme for loop quantum gravity, gr-qc/0305080.

[36] Ashtekar, A. and Lewandowski, L. [2004] Background independent quantum gravity: A status report, *Class. Quant. Grav.* **21**, R53–R152.

[37] Rovelli, C. [2004] *Quantum Gravity* (Cambridge University Press, Cambridge).

[38] Bojowald, M. and Morales-Tecotl, H. A. [2004] Cosmological applications of loop quantum gravity *Lect. Notes Phys.* **646**, 421–462, also available at gr-qc/0306008.

[39] Ambjorn, J., Jurkiewicz, J. and Loll, R. [2004] Emergence of a 4D world from causal quantum gravity, hep-th/0404156.

[40] Gambini, R. and Pullin, J. [2004] Consistent discretizations and quantum gravity, gr-qc/0408025.

[41] Liddle, A. R. and Lyth, D. H. [2000] *Cosmological Inflation and Large Scale Structure* (Cambridge University Press, Cambridge).

[42] Perini, D. [2004] *The Asymptotic Safety Scenario for Gravity and Matter* Ph.D. Dissertation, SISSA.

[43] Thiemann, T. [2005] *Introduction to Modern Canonical Quantum General Relativity* (Cambridge University Press, Cambridge); draft available as gr-qc/0110034.

[44] Ashtekar, A. and Bojowald, M. [2004] Quantum geometry and the Schwarzschild singularity, preprint.

[45] Ashtekar, A. and Bojowald, M. [2005] Black hole evaporation: A paradigm, *Class. Quant. Grav.* **22**, 3349–3362.

[46] Ashtekar, A. [2005] *100 Years of Relativity; Space-time Structure: Einstein and Beyond* (World Scientific, Singapore).

[47] Ashtekar, A., Pawlowski, T. and Singh, P. [2006] Quantum nature of the big bang, *Phys. Rev. Lett.* **96**, 141301–141304; [2006] Quantum nature of the big-bang: an analytical and numerical investigation, *Phys. Rev. D* **73**, 124038–124071; Quantum nature of the big-bang: improved dynamics, *Phys. Rev. D* (at press), gr-qc/0607039.

CHAPTER 6

Brownian Functionals in Physics and Computer Science

****

SATYA N. MAJUMDAR

Laboratoire de Physique Théorique et Modèles Statistiques, Université Paris-Sud. Bât. 100, 91405 Orsay Cedex, France

This is a brief review on Brownian functionals in one dimension and their various applications. After a brief description of Einstein's original derivation of the diffusion equation, this article provides a pedagogical introduction to the path integral methods leading to the derivation of the celebrated Feynman–Kac formula. The usefulness of this technique in calculating the statistical properties of Brownian functionals is illustrated with several examples in physics and probability theory, with particular emphasis on applications in computer science. The statistical properties of "first-passage Brownian functionals" and their applications are also discussed.

1. Introduction

The year 2005 marks the centenary of the publication of three remarkable papers by Einstein, one on Brownian motion [1], one on special relativity [2], and the other one on the photoelectric effect and light quanta [3]. Each of them made a revolution on its own. In particular, his paper on Brownian motion (along with the related work by Smoluchowsky [4] and Langevin [5]) had a more sustained and broader impact, not just in traditional 'natural' sciences such as physics, astronomy, chemistry, biology and mathematics but even in 'man-made' subjects such as economics and computer science. The range of applications of Einstein's Brownian motion and his theory of diffusion is truly remarkable. The ever emerging new applications in diverse fields have made the Brownian motion a true legacy and a great gift of Einstein to science.

There have been numerous articles in the past detailing the history of Brownian motion prior to and after Einstein. Reviewing this gigantic amount of work is beyond the scope of this article. This year two excellent reviews on the Brownian motion with its history and applications have been published, one by Frey and Kroy [6] and the other by Duplantier [7]. The former discusses the applications of Brownian motion in soft matter and biological physics and the latter, after a very nice historical review, discusses the applications of Brownian motion in a variety of two-dimensional growth problems and their connections to the conformal field theory. Apart from these two reviews, there have been numerous other recent reviews on the 100 years of Brownian motion [8] — it is simply not possible to cite all of them within the limited scope of this article and I apologise for that. The purpose of the present article is to discuss some complementary aspects of Brownian motion that are not covered by the recent reviews mentioned above.

After a brief introduction to Einstein's original derivation of the Stokes–Einstein relation and the diffusion equation in Sec. 2, the principal focus of the remainder of the article will be on the statistical properties of functionals of one-dimensional Brownian motion, with special emphasis on their applications in physics and computer science. If $x(\tau)$ represents a Brownian motion, a Brownian functional over a fixed time interval $[0, t]$ is simply defined as $T = \int_0^t U(x(\tau))\, d\tau$, where $U(x)$ is some prescribed arbitrary function. For each realization of the Brownian path, the quantity T has a different value and one is interested in the probability density function (pdf) of T. It was Kac who first realized [9] that the statistical properties of one-dimensional Brownian functionals can be studied by cleverly using the path integral method devised by Feynman in his unpublished Ph.D thesis at Princeton. This observation of Kac thus took Einstein's classical diffusion process into yet another completely different domain of physics namely the quantum mechanics and led to the discovery of the celebrated Feynman–Kac formula. Since then Brownian functionals have found numerous applications in diverse fields ranging from probability theory [9, 10] and finance [11] to disordered systems and mesocopic physics [12]. In this article I will discuss some of them, along with some recent applications of Brownian functionals in computer science.

After a brief and pedagogical derivation of the path integral methods leading to the Feynman–Kac formula in Sec. 3, I will discuss several applications from physics, computer science and graph theory in Sec. 4. In Sec. 5, the statistical properties of "first-passage Brownian functionals" will be discusssed. A first-passage functional is defined as $T = \int_0^{t_f} U(x(\tau))\, d\tau$ where t_f is the first-passage time of the Brownian process $x(\tau)$, i.e. the first time the process crosses zero. Such first-passage functionals have many applications, e.g. in the distribution of lifetimes of comets, in queueing theory and also in transport properties in disordered systems. Some of these applications will be discussed in Sec. 5.

The diverse and ever emerging new applications of Brownian functionals briefly presented here will hopefully convince the reader that 'Brownian functionalogy' merits the status of a subfield of statistical physics (and stochastic calculus) itself and is certainly a part of the legacy that Einstein left behind.

2. Einstein's Theory of Brownian Motion and Langevin's Stochastic Equation

Einstein's 1905 paper on Brownian motion [1] achieved two important milestones: (i) to relate macroscopic kinetic parameters such as the diffusion constant and friction coefficient to the correlation functions characterizing fluctuations of microscopic variables — known as a fluctuation–dissipation relation and (ii) to provide a derivation of the celebrated diffusion equation starting from the microscopic irregular motion of a particle — thus laying the foundation of the important field of "stochastic processes".

2.1. *A fluctuation–dissipation relation*

Very briefly, Einstein's argument leading to the derivation of fluctuation-dissipation relation goes as follows. Imagine a dilute gas of noninteracting Brownian particles in a solvent under a constant volume force K (such as gravity) on each particle. For simplicity, we consider a one-dimensional system here, though the arguments can be generalized straightforwardly to higher dimensions. There are two steps to the argument. The first step is to assume that the dilute gas of Brownian particles suspended in a solvent behaves as an ideal gas and hence exerts an osmotic pressure on the container giving rise to a pressure field. The pressure $p(x)$ at point x is related to the density $\rho(x)$ via the equation of state for an ideal gas: $p(x) = k_B T\rho(x)$, where k_B is the Boltzmann's constant and T is the temperature. The force per unit volume due to the pressure field $-\partial_x p(x)$ must be balanced at equilibrium by the net external force density $K\rho(x)$, leading to the force balance condition: $K\rho(x) = -\partial_x p(x) = -k_B T\partial_x \rho(x)$. The solution is simply

$$\rho(x) = \rho(0) \exp\left(-\frac{K}{k_B T} x\right). \tag{1}$$

The next step of the argument consists of identifying two currents in the system. The first is the diffusion current $j_{\text{diff}} = -D\partial_x \rho(x)$ where D is defined as the diffusion coefficient. The second is the drift current due to the external force, j_{drift} which can be computed as follows. Under a constant external force, each particle achieves at long times a terminal drift velocity, $v = K/\Gamma$ where Γ is the friction coefficient. For spherical particles of radius a, Γ is given by the Stoke's formula, $\Gamma = 6\pi\eta a$ where η is the viscosity. Thus, $j_{\text{drift}} = v\rho(x) = K\rho(x)/\Gamma$. Now, at equilibrium, the net current in a closed system must be zero, $j = j_{\text{diff}} + j_{\text{drift}} = 0$ leading to the equation $-D\partial_x \rho(x) + K\rho(x)/\Gamma = 0$. The solution is

$$\rho(x) = \rho(0) \exp\left(-\frac{K}{\Gamma D} x\right). \tag{2}$$

Comparing Eqs. (1) and (2) Einstein obtained the important relation

$$D = \frac{k_B T}{\Gamma}, \tag{3}$$

which is known today as the Stokes–Einstein relation that connects macroscopic kinetic coefficients such as D and Γ to the thermal fluctuations characterized by the temperature T.

2.2. *Diffusion as a microscopic process*

In addition to the fluctuation–dissipation relation in Eq. (3), Einstein's 1905 paper on Brownian motion also provided an elegant derivation of the diffusion equation that expressed the diffusion constant D in terms of microscopic fluctuations. Since the particles are independent, the density $\rho(x,t)$ can also be interpreted as the probability $\rho(x,t) \equiv P(x,t)$ that a single Brownian particle is at position x at time t and the aim is to derive an evolution equation for $P(x,t)$ by following the trajectory of a single particle. Here one assumes that the particle is free, i.e. not subjected to any external drift. Einstein considered the particle at position x at time t and asssumed that in a microscopic time step Δt, the particle jumps by a random amount Δx which is thus a stochastic variable. He then wrote down an evolution equation for $P(x,t)$

$$P(x,t+\Delta t) = \int_{-\infty}^{\infty} P(x-\Delta x, t)\phi_{\Delta t}(\Delta x)\, d(\Delta x) \tag{4}$$

where $\phi_{\Delta t}(\Delta x)$ is the normalized probability density of the 'jump' Δx in time step Δt. This evolution equation is known today as the Chapman–Kolmogorov equation and it inherently assumes that the stochastic process $x(t)$ is Markovian. This means that the jump variables Δx's are independent from step to step, so that the position $x(t)$ of the particle at a given time step depends only on its previous time step and not on the full previous history of evolution. Next Einstein assumed that $P(x-\Delta x, t)$ in the integrand in Eq. (4) can be Taylor expanded assuming 'small' Δx. This gives

$$P(x,t+\Delta t) = P(x,t) - \mu_1 \frac{\partial P}{\partial x} + \frac{\mu_2}{2!}\frac{\partial^2 P}{\partial x^2} + \dots \tag{5}$$

where $\mu_k = \int_{-\infty}^{\infty} (\Delta x)^k \phi_{\Delta t}(\Delta x) d(\Delta x)$ is the k-th moment of the jump variable Δx. Furthermore, the absence of external drift sets $\mu_1 = 0$. Dividing both sides of Eq. (5) by Δt, taking the limit $\Delta t \to 0$ and keeping only the leading nonzero term (assuming the higher order terms vanish as $\Delta t \to 0$) one gets the diffusion equation

$$\frac{\partial P}{\partial t} = D\frac{\partial^2 P}{\partial x^2} \tag{6}$$

where the diffusion constant

$$D = \lim_{\Delta t \to 0} \frac{\mu_2}{2\Delta t} = \lim_{\Delta t \to 0} \frac{1}{2\Delta t}\int_{-\infty}^{\infty} (\Delta x)^2 \phi_{\Delta t}(\Delta x) d(\Delta x) = \lim_{\Delta t \to 0} \frac{\langle (\Delta x)^2 \rangle}{2\,\Delta t}, \tag{7}$$

where $\langle (\Delta x)^2 \rangle$ is the average of the square of the microscopic displacement in a microscopic time step Δt. Thus Einstein was able to express the constant D that appears as a coefficient in the macrosopic diffusion current $j_{\text{diff}} = -D\partial_x P$ in terms of the microscopic fluctuation Δx in the position of a Brownian particle. This derivation also brings out the fundamental principle of the diffusion process, i.e. the length scale must scale as the square root of the time scale.

The position of the Brownian particle can evolve via many possible 'stochastic' trajectories. The diffusion equation (6) describing the evolution of the probability density sums up the effects of all underlying stochastic trajectories. However, it is often useful to have a mathematical description of each single trajectory. This brings us to the description of the diffusion process à la Langevin [5]. It is clear from Einstein's derivation that the local slope of an evolving trajectory at time t can be written as

$$\frac{\Delta x}{\Delta t} = \xi_{\Delta t}(t) \tag{8}$$

where $\xi_{\Delta t}(t)$ is a random 'noise' which is independent from one microscopic step to another, and it has zero mean. Its variance at a given time t, in the continuum limit $\Delta t \to 0$, can also be computed from Eq. (7). One gets $\langle \xi_{\Delta t}^2(t)\rangle = \langle (\Delta x)^2\rangle/(\Delta t)^2 = 2D/\Delta t$ as $\Delta t \to 0$. Thus the noise term typically scales as $1/\sqrt{\Delta t}$ as $\Delta t \to 0$. The correlation function of the noise between two different times can then be written as,

$$\begin{aligned} \langle \xi_{\Delta t}(t)\xi_{\Delta t}(t')\rangle &= 0 \qquad \text{if } t \neq t' \\ &= \frac{2D}{\Delta t} \qquad \text{if } t = t' . \end{aligned} \tag{9}$$

In the continuum limit $\Delta t \to 0$, the noise $\xi_{\Delta t}(t)$ then tends to a limiting noise $\xi(t)$ which has zero mean and a correlator, $\langle \xi(t)\xi(t')\rangle = 2D\delta(t-t')$. This last result follows by formally taking the limit $\Delta t \to 0$ in Eq. (9) where, loosely speaking, one replaces the $1/\Delta t$ by $\delta(0)$. Such a noise is called a 'white' noise. Thus, in the continuum limit $\Delta t \to 0$, Eq. (8) reduces to the celebrated Langevin equation,

$$\frac{dx}{dt} = \xi(t) \tag{10}$$

where $\xi(t)$ is a white noise. Moreover, in the continuum limit $\Delta t \to 0$, one can assume, without any loss of generality, that the white noise $\xi(t)$ is Gaussian. This means that the joint probability distribution of a particular history of the noise variables $[\{\xi(\tau)\}, \text{for } 0 \leq \tau \leq t]$ can be written as

$$\text{Prob}\,[\{\xi(\tau)\}] \propto \exp\left[-\frac{1}{4D}\int_0^t \xi^2(\tau)d\tau\right] . \tag{11}$$

We will see later that this particular fact plays the key role in the representation of Brownian motion as a path integral. The Brownian motion $x(t)$ can thus be represented as the integrated white noise, $x(t) = x(0) + \int_0^t \xi(\tau)d\tau$. While the physicists call this a Brownian motion, the mathematicians call this integrated white noise the Wiener process, named after the mathematician N. Wiener.

Langevin's formulation in Eq. (10) also makes a correspondence between Brownian motion and the random walk problem where the position x_n of a random walker after n steps evolves via

$$x_n = x_{n-1} + \xi_n \tag{12}$$

where ξ_n's are independent random variables, each drawn from the common distribution $\phi(\xi)$ for each step n. In fact, the idea of understanding Brownian motion in terms of random walks was first conceived by Smoluchowsky [4]. The Langevin equation representation of Brownian motion makes this connection evident, the Brownian motion is just the suitably taken continuum limit of the random walk problem. For large n, by virtue of the central limit theorem, the results for the random walk problem reduce to those of the Brownian motion. This is an important point because in many applications, especially those in computer science as will be discussed later, one often encounters discrete random walks as in Eq. (12) which are often more difficult to solve than the continuum Brownian motion. However, since in most applications one is typically interested in the large time scaling-limit results, one can correctly approximate a discrete random walk sequence by the continuum Brownian process and this makes life much simpler.

3. Brownian Process as a Path Integral

The solution of the diffusion equation (6) can be easily obtained in free space by the Fourier transform method. For simplicity, we set $D = 1/2$ for the rest of the article. One gets

$$P(x,t) = \int_{-\infty}^{\infty} dx_0\, G_0(x,t|x_0,0)\, P(x_0,0) \tag{13}$$

where $P(x_0,0)$ is the initial condition and the diffusion propagator

$$G_0(x,t|x_0,0) = \frac{1}{\sqrt{2\pi t}} \exp\left[-(x-x_0)^2/2t\right] \tag{14}$$

denotes the conditional probability that the Brownian particle reaches x at time t, starting from x_0 at $t = 0$. It was M. Kac who first made the important observation [9] that this diffusion propagator can be interpreted, using Feynman's path integral formalism, as the quantum propagator of a free particle from time 0 to time t. This is easy to see. Using the property of the Gaussian noise in Eq. (11) and the Langevin equation (10), it is clear that the probability of any path $\{x(\tau)\}$ can be written as

$$P[\{x(\tau)\}] \propto \exp\left[-\frac{1}{2}\int_0^t \left(\frac{dx}{d\tau}\right)^2 d\tau\right]. \tag{15}$$

Thus the diffusion propagator, i.e. the probability that a path goes from x_0 at $t=0$ to x at t can be written as a sum of the contributions from all possible paths propagating from x_0 at $\tau=0$ to x at $\tau=t$. This sum is indeed Feynman's path integral [13]

$$G_0(x,t|x_0,0) = \int_{x(0)=x_0}^{x(t)=x} \mathcal{D}x(\tau) \exp\left[-\frac{1}{2}\int_0^t \left(\frac{dx}{d\tau}\right)^2 d\tau\right]. \tag{16}$$

One immediately identifies the term $\frac{1}{2}\left(\frac{dx}{d\tau}\right)^2$ as the classical kinetic energy of a particle of unit mass and the integral $\frac{1}{2}\int_0^t \left(\frac{dx}{d\tau}\right)^2 d\tau$ as the Lagrangian of a free particle of unit mass. Following Feynman [13], one then identifies the path integral in Eq. (16) as a quantum propagator

$$G_0(x,t|x_0,0) = \langle x|e^{-\hat{H}_0 t}|x_0\rangle \tag{17}$$

where $\hat{H}_0 \equiv -\frac{1}{2}\frac{\partial^2}{\partial x^2}$ is the quantum Hamiltonian of a free particle (we have set the mass $m=1$ and the Planck's constant $\hbar=1$). To make the connection complete, the quantum propagator on the r.h.s. of Eq. (17) can be easily evaluated by expanding it in the free particle eigenbasis. Noting that $\hat{H}_0$ has free particle eigenfunctions $\psi_k(x) = \frac{1}{\sqrt{2\pi}}e^{ikx}$ with eigenvalue $k^2/2$, one gets

$$\begin{aligned} G_0(x,t|x_0,0) = \langle x|e^{-\hat{H}_0 t}|x_0\rangle &= \int_{-\infty}^{\infty} \langle x|k\rangle\langle k|x_0\rangle e^{-k^2t/2}dk \\ &= \frac{1}{2\pi}\int_{-\infty}^{\infty} e^{ik(x-x_0)-k^2t/2}dk\,. \end{aligned} \tag{18}$$

Performing the Gaussian integration, one gets back the classical result in Eq. (14) that was obtained by solving the diffusion equation. Thus the two approaches, one by solving a partial differential equation usually referred to as the Fokker–Planck approach and the other using the path integral method are completely equivalent.

One may argue that once the basic propagator is known, the Brownian motion is well understood and there is nothing else interesting left to study! This is simply not true because there are intricate questions associated with the diffusion process that are often rather nontrivial. A notable nontrivial example is the calculation of the persistence exponent associated with a diffusion process [14]. Consider a diffusive field $\phi(\vec{r},t)$ evolving via the d-dimensional diffusion equation

$$\frac{\partial \phi}{\partial t} = \nabla^2 \phi \tag{19}$$

starting from the initial condition $\phi(\vec{r},0)$ which is a random Gaussian field, uncorrelated in space. The solution at time t can be easily found using d-dimensional trivial generalization of the diffusion propagator in Eq. (14)

$$\phi(\vec{r},t) = \frac{1}{(2\pi t)^{d/2}} \int d\vec{r}_0 \phi(\vec{r}_0,0) \exp\left[-(\vec{r}-\vec{r}_0)^2/2t\right] . \tag{20}$$

Now, suppose that we fix a point $\vec{r}$ in space and monitor the field $\phi(\vec{r},t)$ there as a function of time t and ask: what is the probability $P(t)$ that the field $\phi(\vec{r},t)$ at $\vec{r}$ does not change sign up to time t starting initially at the random value $\phi(\vec{r},0)$? By translational invariance, $P(t)$ does not depend on the position $\vec{r}$. This probability $P(t)$ is called the persistence probability that has generated a lot of interest over the last decade in the context of nonequilibrium systems [14]. For the simple diffusion process in Eq. (20), it is known, both theoretically [15] and experimentally [16] that at late times t, the persistence $P(t)$ has a power law tail $P(t) \sim t^{-\theta}$ where the persistence exponent θ is nontrivial (even in one dimension!), e.g. $\theta \approx 0.1207$ in $d = 1$, $\theta \approx 0.1875$ in $d = 2$, $\theta \approx 0.2380$ in $d = 3$, etc. While this exponent θ is known numerically very precisely and also very accurately by approximate analytical methods [15], an exact calculation of θ has not yet been achieved and it remains as an outstanding unsolved problem for the diffusion process [17]. This example thus clarifies that while the knowledge of the diffusion propagator is necessary, it is by no means sufficient to answer more detailed history related questions associated with the diffusion process.

Note that in the persistence problem discussed above, the relevant stochastic process at a fixed point $\vec{r}$ in space, whose properties one is interested in, is actually a more complex non-Markovian process [14] even though it originated from a simple diffusion equation. In this article, we will stay with our simple Brownian motion in Eq. (10) which is a Markov process and discuss some of the nontrivial aspects of this simple Brownian motion. For example, in many applications of Brownian motion in physics, finance and computer science, the relevant Brownian process is often constrained. For example, an important issue is the first-passage property of a Brownian motion [18–20], i.e. the distribution of the first time that a Brownian process crosses the origin? For this, one needs to sample only a subset of all possible Brownian paths that do not cross the origin up to a certain time. This can be achieved by imposing the constraint of no crossing on a Brownian path. Apart from the constrained Brownian motion, some other applications require a knowledge of the statistical properties of a Brownian functional up to time t, defined as $T_t = \int_0^t U(x(\tau))\, d\tau$, where $U(x)$ is a specified function. We will provide several examples later and will see that while the properties of a free Brownian motion are rather simple and are essentially encoded in its propagator in Eq. (14), properties of constrained Brownian motion or that of a Brownian functional are often nontrivial to derive and the path integral technique discussed above is particularly suitable to address some of these issues.

3.1. *Brownian motion with constraints: first-passage property*

As a simple example of a constrained Brownian motion, we calculate in this subsection the first-passage probability density $f(x_0, t)$. The quantity $f(x_0, t)dt$ is simply the probability that a Brownian path, starting at x_0 at $t = 0$, will cross the origin for the first time between time t and $t + dt$. Clearly, $f(x_0, t) = -dq(x_0, t)/dt$ where $q(x_0, t)$ is the probability that the path starting at x_0 at $t = 0$ does not cross the origin up to t. The probability $q(x_0, t)$ can be easily expressed in terms of a path integral

$$q(x_0, t) = \int_0^\infty dx \int_{x(0)=x_0}^{x(t)=x} \mathcal{D}x(\tau) \exp\left[-\frac{1}{2}\int_0^t \left(\frac{dx}{d\tau}\right)^2 d\tau\right] \prod_{\tau=0}^{t} \theta[x(\tau)] \quad (21)$$

where the paths propagate from the initial position $x(0) = x_0$ to the final position x at time t and then we integrate x over only the positive half-space since the final position x can only be positive. The term $\prod_{\tau=0}^t \theta[x(\tau)]$ inside the path integral is an indicator function that enforces the constraint that the path stays above the origin up to t. We then identify the path integral in Eq. (21) as an integral over a quantum propagator,

$$q(x_0, t) = \int_0^\infty dx G(x, t|x_0, 0); \qquad G(x, t|x_0, 0) = \langle x|e^{-\hat{H}_1 t}|x_0\rangle \quad (22)$$

where the Hamiltonian $\hat{H}_1 \equiv -\frac{1}{2}\frac{\partial^2}{\partial x^2} + V(x)$ with the quantum potential $V(x) = 0$ if $x > 0$ and $V(x) = \infty$ if $x \le 0$. The infinite potential for $x \le 0$ takes care of the constraint that the path cannot cross the origin, i.e. it enforces the condition $\prod_{\tau=0}^t \theta[x(\tau)]$. The eigenfunction of $\hat{H}_1$ must vanish at $x = 0$, but for $x > 0$ it corresponds to that of a free particle. The correctly normalized eigenfunctions are thus $\psi_k(x) = \sqrt{\frac{2}{\pi}}\sin(kx)$ with $k \ge 0$ with eigenvalues $k^2/2$. The quantum propagator can then be evaluated again by decomposing into the eigenbasis

$$G(x,t|x_0,0) = \frac{2}{\pi}\int_0^\infty \sin(kx_0)\sin(kx)e^{-k^2t/2}dk$$
$$= \frac{1}{\sqrt{2\pi t}}\left[e^{-(x-x_0)^2/2t} - e^{-(x+x_0)^2/2t}\right]. \quad (23)$$

Note that this result for the propagator can also be derived alternately by solving the diffusion equation with an absorbing boundary condition at the origin. The result in Eq. (23) then follows by a simple application of the image method [18, 20]. Integrating over the final position in x one gets from Eq. (22) the classical result [19], $q(x_0,t) = \mathrm{erf}(x/\sqrt{2t})$ where $\mathrm{erf}(z) = \frac{2}{\sqrt{\pi}}\int_0^z e^{-u^2}du$. The first-passage probability density is then given by

$$f(x_0,t) = -\frac{dq(x_0,t)}{dt} = \frac{x_0}{\sqrt{2\pi}}\frac{e^{-x_0^2/2t}}{t^{3/2}}. \quad (24)$$

For $t \gg x_0^2$, one thus recovers the well known $t^{-3/2}$ decay of the first-passage probability [18, 19] density.

3.2. *Brownian functionals: Feynman–Kac formula*

In this subsection we will discuss how to calculate the statistical properties of a Brownian functional defined as

$$T = \int_0^t U(x(\tau))d\tau \quad (25)$$

where $x(\tau)$ is a Brownian path starting from x_0 at $\tau = 0$ and propagating up to time $\tau = t$ and $U(x)$ is a specified function. Clearly T is random variable taking different values for different Brownian paths. The goal is to calculate its probability distribution $P(T,t|x_0)$. The choice of $U(x)$ depends on which quantity we want to calculate. Brownian functionals appear in a wide range of problems across different fields ranging from probability theory, finance, data analysis, disordered systems and computer science. We consider a few examples below.

1. In probability theory, an important object of interest is the occupation time, i.e. the time spent by a Brownian motion above the origin within a time window of size t [21]. Thus the occupation time is simply, $T = \int_0^t \theta[x(\tau)]d\tau$. Thus, in this problem the function $U(x) = \theta(x)$.

2. For fluctuating $(1 + 1)$-dimensional interfaces of the Edwards–Wilkinson [22] or the Kardar–Parisi–Zhang (KPZ) [23] varieties, the interface profile in the stationary state is described by a one-dimensional Brownian motion in space [24]. The fluctuations in the stationary state are captured by the pdf of the spatially averaged variance of height fluctuations [25] in a finite system of size L, i.e. the pdf of $\sigma^2 = \frac{1}{L}\int_0^L h^2(x)dx$ where $h(x)$ is the deviation of the height from its spatial average. Since $h(x)$ performs a Brownian motion in space, σ^2 is a functional of the Brownian motion as in Eq. (25) with $U(x) = x^2$.
3. In finance, a typical stock price $S(\tau)$ is sometimes modeled by the exponential of a Brownian motion, $S(\tau) = e^{-\beta x(\tau)}$, where β is a constant. An object that often plays a crucial role is the integrated stock price up to some 'target' time t, i.e. $T = \int_0^t e^{-\beta x(\tau)} d\tau$ [26]. Thus in this problem $U(x) = e^{-\beta x}$. Interestingly, this integrated stock price has an interesting analogy in a disordered system where a single overdamped particle moves in a random potential. A popular model is the so-called Sinai model [27] where the random potential is modeled as the trace of a random walker in space. Interpreting the time τ as the spatial distance, $x(\tau)$ is then the potential energy of the particle and $e^{-\beta x(\tau)}$ is just the Boltzmann factor. The total time t is just the size of a linear box in which the particle is moving. Thus $T = \int_0^t e^{-\beta x(\tau)} d\tau$ is just the partition function of the particle in a random potential [28]. In addition, the exponential of a Brownian motion also appears in the expression for the Wigner time delay in one-dimensional quantum scattering process by a random potential [29].
4. In simple models describing the stochastic behavior of daily temperature records, one assumes that the daily temperature deviation from its average is a simple Brownian motion $x(\tau)$ in a harmonic potential (the Ornstein–Uhlenbeck process). Then the relevant quantity whose statistical properties are of interest is the so-called 'heating degree days' (HDD) defined as $T = \int_0^t x(\tau)\,\theta(x(\tau))\,d\tau$ that measures the integrated excess temperature up to time t [30]. Thus in this example, the function $U(x) = x\theta(x)$.
5. Another quantity, first studied in the context of economics [31] and later extensively by probabilists [32] is the total area (unsigned) under a Brownian motion, i.e. $T = \int_0^t |x(\tau)|\,d\tau$. Thus in this example, $U(x) = |x|$. The same functional was also studied by physicists in the context of electron-electron and phase coherence in one-dimensional weakly disordered quantum wire [33].

We will mention several other examples as we go along. Note that in all the examples mentioned above the function $U(x)$ is such that the random variable T has only positive support. Henceforth we will assume that. For a given such function $U(x)$, how does one calculate the pdf of T? It was Kac who, using the path integral techniques developed by Feynman in his Ph.D thesis, first devised a way of computing the pdf $P(T,t|x_0)$ of a Brownian functional [9] that led to the famous Feynman–Kac formula. We summarize below Kac's formalism.

Feynman–Kac formula: Since T has only positive support, a natural step is to introduce the Laplace transform of the pdf $P(T,t|x_0)$,

$$Q(x_0,t) = \int_0^\infty e^{-pT} P(T,t|x_0)dT = E_{x_0}[e^{-p\int_0^t U(x(\tau))d\tau}] \tag{26}$$

where the r.h.s is an expectation over all possible Brownian paths $\{x(\tau)\}$ that start at x_0 at $\tau = 0$ and propagate up to time $\tau = t$. We have, for notational simplicity, suppressed the p dependence of $Q(x_0,t)$. Using the measure of the Brownian path in Eq. (15), one can then express the expectation on the r.h.s of Eq. (26) as a path integral

$$Q(x_0,t) = E_{x_0}[e^{-p\int_0^t U(x(\tau))d\tau}]$$
$$= \int_{-\infty}^\infty dx \int_{x(0)=x_0}^{x(t)=x} \mathcal{D}x(\tau) \exp\left[-\int_0^t d\tau \left[\frac{1}{2}\left(\frac{dx}{d\tau}\right)^2 + pU(x(\tau))\right]\right] \tag{27}$$
$$= \int_{-\infty}^\infty dx \langle x|e^{-\hat{H}t}|x_0\rangle \tag{28}$$

where the quantum Hamiltonian $\hat{H} \equiv -\frac{1}{2}\frac{\partial^2}{\partial x^2} + pU(x)$ corresponds to the Shrödinger operator with a potential $pU(x)$. Note that in Eq. (27) all paths propagate from $x(0) = x_0$ to $x(t) = x$ in time t and then we have integrated over the final position x. The quantum propagator $G(x,t|x_0) = \langle x|e^{-\hat{H}t}|x_0\rangle$ satisfies a Shrödinger-like equation

$$\frac{\partial G}{\partial t} = \frac{1}{2}\frac{\partial^2 G}{\partial x^2} - pU(x)G \tag{29}$$

which can be easily established by differentiating $G(x,t|x_0) = \langle x|e^{-\hat{H}t}|x_0\rangle$ with respect to t and using the explicit representation of the operator $\hat{H}$. The initial condition is simply, $G(x,0|x_0) = \delta(x - x_0)$. Thus the scheme of Kac involves three steps: (i) solve the partial differential equation (29) to get $G(x,t|x_0)$ (ii) integrate $G(x,t|x_0)$ over the final position x as in Eq. (28) to obtain the Laplace transform $Q(x_0,t)$ and (iii) invert the Laplace transform in Eq. (26) to obtain the pdf $P(T,t|x_0)$. Equations (26), (28) and (29) are collectively known as the celebrated Feynman–Kac formula.

A shorter backward Fokker–Planck approach: An alternative and somewhat shorter approach would be to write down a partial differential equation for $Q(x_0,t)$ in Eq. (28) directly. An elementary exercise yields

$$\frac{\partial Q}{\partial t} = \frac{1}{2}\frac{\partial^2 Q}{\partial x_0^2} - p\,U(x_0)\,Q \tag{30}$$

where note that the spatial derivatives are with respect to the *initial* position x_0. This is thus a 'backward' Fokker–Planck approach as opposed to the 'forward' Fokker–Planck equation satisfied by G in Eq. (29) of Kac where the spatial derivatives are with respect to the *final* position of the particle. Basically we have reduced the additional step (ii) of integrating over the final position in Kac's derivation. The solution $Q(x_0,t)$ of Eq. (30) must satisfy the initial condition $Q(x_0,0)=1$ that follows directly from the definition in Eq. (26). To solve Eq. (30), it is useful to take a further Laplace transform of Eq. (30) with respect to t, $\tilde{Q}(x_0,\alpha)=\int_0^\infty Q(x_0,t)e^{-\alpha t}dt$. Using the initial condition $Q(x_0,0)=1$, one arrives at an ordinary second order differential equation

$$\frac{1}{2}\frac{d^2\tilde{Q}}{dx_0^2} - [\alpha + pU(x_0)]\tilde{Q} = -1 \tag{31}$$

which needs to be solved subject to the appropriate boundary conditions that depend on the behavior of the function $U(x)$ at large x. Given that $T=\int_0^t U(x(\tau))\,d\tau$ has positive support, there are two typical representative asymptotic behaviors of $U(x)$:

1. If the function $U(x)$ approaches a constant value at large x, i.e. $U(x)\to c_{\pm}$ as $x\to\pm\infty$, then it is easy to argue (for an example, see below) that $Q(x_0\to\pm\infty,\alpha)=1/[p\,c_{\pm}+\alpha]$. In this case, the underlying quantum Hamiltonian $\hat{H}\equiv -\frac{1}{2}\frac{\partial^2}{\partial x^2}+pU(x)$ has scattering states in its spectrum, in addition to possible bound states.
2. If the function $U(x)\to\infty$ as $x\to\pm\infty$, then $Q(x_0\to\pm\infty,\alpha)=0$. In this case the underlying quantum Hamiltonian $\hat{H}$ has only bound states and hence a discrete spectrum.

Thus, in principle, knowing the solution $\tilde{Q}(x_0,\alpha)$ of Eq. (31), the original pdf $P(T,t|x_0)$ can be obtained by inverting the double Laplace transform

$$\tilde{Q}(x_0,\alpha) = \int_0^\infty dt e^{-\alpha t} \int_0^\infty dT e^{-pT}\, P(T,t|x_0)\,. \tag{32}$$

Below we provide an example where all these steps can be carried out explicitly to obtain an exact closed form expression for the pdf $P(T,t|x_0)$.

3.3. *A simple illustration: Lévy's arcsine law for the distribution of the occupation time*

As an illustration of the method outlined in the previous subsection, let us calculate the distribution of the occupation time $T = \int_0^t \theta[x(\tau)]d\tau$. This distribution was first computed by Lévy using probabilistic methods [21]. Later Kac derived it using Feynman–Kac formalism discussed above [9]. We present here a derivation based on the backward Fokker–Planck approach outlined above.

Substituting $U(x_0) = \theta(x_0)$ in Eq. (31) we solve the differential equation separately for $x_0 > 0$ and $x_0 < 0$ and then match the solution at $x_0 = 0$ by demanding the continuity of the solution and that of its first derivative. In addition, we use the boundary conditions $\tilde{Q}(x_0 \to \infty, \alpha) = 1/(\alpha+p)$ and $\tilde{Q}(x_0 \to -\infty, \alpha) = 1/\alpha$. They follow from the observations:

1. If the starting point $x_0 \to \infty$, the particle will stay on the positive side for all finite t implying $T = \int_0^t \theta(x(\tau))\, d\tau = t$ and hence $Q(x_0 \to \infty, t) = E[e^{-pT}] = e^{-pt}$ and its Laplace transform $\tilde{Q}(x_0 \to \infty, \alpha) = \int_0^\infty e^{-(\alpha+p)\,t} dt = 1/(\alpha+p)$.
2. If the starting point $x_0 \to -\infty$, the particle stays on the negative side up to any finite t implying $T = \int_0^t \theta(x(\tau))\, d\tau = 0$ and hence $Q(x_0 \to -\infty, t) = E[e^{-pT}] = 1$ and its Laplace transform $\tilde{Q}(x_0 \to -\infty, \alpha) = \int_0^\infty e^{-\alpha t} dt = 1/\alpha$.

Using these boundary and matching conditions, one obtains an explicit solution

$$\tilde{Q}(x_0,\alpha) = \frac{1}{(\alpha+p)}\left[1 + \frac{(\sqrt{\alpha+p}-\sqrt{\alpha})}{\sqrt{\alpha}}\, e^{-\sqrt{2(\alpha+p)}\,x_0}\right] \quad \text{for } x_0 > 0 \tag{33}$$

$$= \frac{1}{\alpha}\left[1 + \frac{(\sqrt{\alpha}-\sqrt{\alpha+p})}{\sqrt{\alpha+p}}\, e^{\sqrt{2\alpha}\,x_0}\right] \quad \text{for } x_0 < 0\,. \tag{34}$$

The solution is simpler if the particle starts at the origin $x_0 = 0$. Then one gets from above

$$\tilde{Q}(0,\alpha) = \frac{1}{\sqrt{\alpha(\alpha+p)}}. \tag{35}$$

Inverting the Laplace transform, first with respect to p and then with respect to α, one obtains the pdf of the occupation time for all $0 \le T \le t$

$$P(T,t|x_0=0) = \frac{1}{\pi}\frac{1}{\sqrt{T(t-T)}}. \tag{36}$$

In particular, the cumulative distribution

$$\int_0^T P(T',t|x_0=0)dT' = \frac{2}{\pi}\arcsin\left(\sqrt{\frac{T}{t}}\right) \tag{37}$$

is known as the famous arcsine law of Lévy [21].

The result in Eq. (36) is interesting and somewhat counterintuitive. The probability density peaks at the two end points $T = 0$ and $T = t$ and has a minimum at $T = 1/2$ which is also the average occupation time. Normally one would expect that any 'typical' path would spend roughly half the time $t/2$ on the positive side and the other half on the negative side. If that was the case, one would have a peak of the occupation time distribution at the average value $t/2$. The actual result is exactly the opposite — one has a minimum at $T = t/2$! This means that a typical path, starting at the origin, tends to stay either entirely on the positive side (explaining the peak at $T = t$) or entirely on the negative side (explaining the peak at $T = 0$). In other words, a typical Brownian path is 'stiff' and reluctant to cross the origin. This property that 'the typical is not the same as the average' is one of the hidden surprises of Einstein's Brownian motion.

The concept of the occupation time and related quantities have been studied by probabilists for a long time [34]. Recently they have played important roles in physics as well, for example, in understanding the dynamics out of equilibrium in coarsening systems [35], ergodicity properties in anomalously diffusive processses [36], in renewal processes [37], in models related to spin glasses [38], in understanding certain aspects of transport properties in disordered sysyems [39] and also in simple models of blinking quantum dots [40].

4. Area Under a Brownian Excursion: Applications in Physics and Computer Science

In this section we consider an example where, by applying the path integral method outlined in the previous section, one can compute exactly the distribution of a functional of a Brownian process that is also constrained to stay positive over a fixed time interval $[0, t]$. A Brownian motion $x(\tau)$ in an interval $0 \leq \tau \leq t$, that starts and ends at the origin $x(0) = x(t) = 0$ but is conditioned to stay positive in between, is called a Brownian excursion. The area under the excursion, $A = \int_0^T x(\tau)d\tau$, is clearly a random variable taking a different value for each realization of the excursion. A natural question that the mathematicians have studied quite extensively [41–45] over the past two decades is: what is the pdf $P(A, t)$ of the area under a Brownian excursion over the interval $[0, t]$? Since the typical lateral displacement of the excursion at time τ scales as $\sqrt{\tau}$, it follows that the area over the interval $[0, t]$ will scale as $t^{3/2}$ and hence its pdf must have a scaling form, $P(A, t) = t^{-3/2} f(A/t^{3/2})$. The normalization condition $\int_0^\infty P(A, t)dA = 1$ demands a prefactor $t^{-3/2}$ and also the conditions: $f(x) \geq 0$ for all x and $\int_0^\infty f(x)dx = 1$. One then interprets the scaling function $f(x)$ as the distribution of the area under the Brownian excursion $x(u)$ over a *unit* interval $u \in [0, 1]$. The function $f(x)$, or rather its Laplace transform, was first computed analytically by Darling [41] and independently by Louchard [42],

$$\tilde{f}(s) = \int_0^\infty f(x)e^{-sx}dx = s\sqrt{2\pi} \sum_{k=1}^{\infty} e^{-\alpha_k s^{2/3} 2^{-1/3}}, \tag{38}$$

where α_k's are the magnitudes of the zeros of the standard Airy function $\mathrm{Ai}(z)$. The Airy function $\mathrm{Ai}(z)$ has discrete zeros on the negative real axis at e.g. $z = -2.3381$, $z = -4.0879$, $z = -5.5205$, etc. Thus, $\alpha_1 = 2.3381\ldots$, $\alpha_2 = 4.0879\ldots$, $\alpha_3 = 5.5205\ldots$, etc. Since the expression of $f(x)$ involves the zeros of Airy function, the function $f(x)$ has been named the Airy distribution function [44], which should not be confused with the Airy function $\mathrm{Ai}(x)$ itself. Even though Eq. (38) provides a formally exact expression of the Laplace transform, it turns out that the calculation of the moments $M_n = \int_0^\infty x^n f(x)dx$ is highly nontrivial and they can be determined only recursively [43] (see Sec. 2). Takacs was able to formally invert the Laplace transform in Eq. (38) to obtain [43],

$$f(x) = \frac{2\sqrt{6}}{x^{10/3}} \sum_{k=1}^{\infty} e^{-b_k/x^2} b_k^{2/3} U(-5/6, 4/3, b_k/x^2), \tag{39}$$

where $b_k = 2\alpha_k^3/27$ and $U(a,b,z)$ is the confluent hypergeometric function [47]. The function $f(x)$ has the following asymptotic tails [43, 48],

$$\begin{aligned} f(x) &\sim x^{-5}\, e^{-2\alpha_1^3/27x^2} \quad \text{as} \quad x \to 0 \\ f(x) &\sim e^{-6x^2} \quad \text{as} \quad x \to \infty \,. \end{aligned} \tag{40}$$

So, why would anyone care about such a complicated function? The reason behind the sustained interest and study [43–46] of this function $f(x)$ seems to be the fact that it keeps resurfacing in a number of seemingly unrelated problems, in computer science, graph theory, two-dimensional growth problems and more recently in fluctuating one-dimensional interfaces. We discuss below some of these applications.

The result in Eq. (38) was originally derived using probabilistic methods [41, 42]. A more direct physical derivation using the path integral method was provided more recently [49], which we outline below. Following the discussion in the previous section, our interest here is in the functional $T = A = \int_0^t x(\tau)d\tau$. However, we also need to impose the constraint that the path stays positive between 0 and t, i.e. we have to insert a factor $\prod_\tau \theta[x(\tau)]$ in the path integral. However, one needs to be a bit careful in implementing this constraint. Note that the path starts at the origin, i.e. $x(0) = 0$. But if we take a continuous time Brownian path that starts at the origin, it immediately recrosses the origin many times and hence it is impossible to restrict a Brownian path to be positive over an interval if it starts at the origin. One can circumvent this problem by introducing a small cut-off ϵ, i.e. we consider all paths that start at $x(0) = \epsilon$ and end at $x(t) = \epsilon$ and stays positive in between (see Fig. 1). We then first derive the pdf $P(A, t, \epsilon)$ and then take the limit $\epsilon \to 0$ eventually.

Following the method in the previous section, the Laplace transform of the pdf is now given by

$$\begin{aligned} Q(\epsilon, t) &= E_\epsilon\left[e^{-p\int_0^t x(\tau)d\tau}\right] \\ &= \frac{1}{Z_E}\int_{x(0)=\epsilon}^{x(t)=\epsilon} \mathcal{D}x(\tau) e^{-\int_0^t d\tau\left[\frac{1}{2}(dx/d\tau)^2 + px(\tau)\right]} \prod_{\tau=0}^{t} \theta[x(\tau)] \,. \end{aligned} \tag{41}$$

where Z_E is a normalization constant

$$Z_E = \int_{x(0)=\epsilon}^{x(t)=\epsilon} \mathcal{D}x(\tau)\, e^{-\frac{1}{2}\int_0^t d\tau\,(dx/d\tau)^2} \prod_{\tau=0}^{t} \theta[x(\tau)] \tag{42}$$

that is just the partition function of the Brownian excursion.

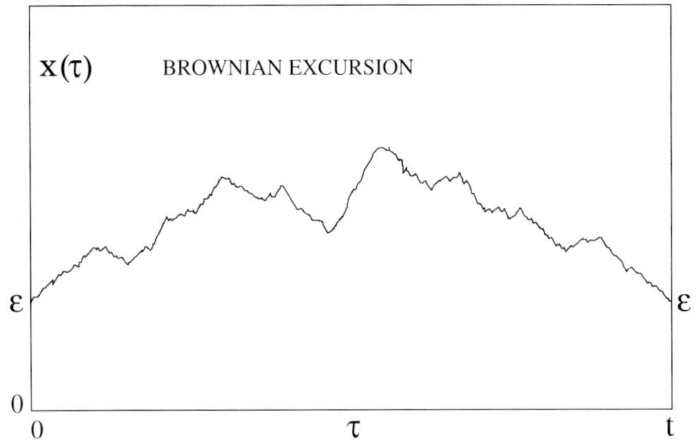

Fig. 1. A Brownian excursion over the time interval $0 \leq \tau \leq t$ starting at $x(0) = \epsilon$ and ending at $x(t) = \epsilon$ and staying positive in between.

Clearly, $Z_E = \langle \epsilon | e^{-\hat{H}_1 t} | \epsilon \rangle$ where $\hat{H}_1 \equiv -\frac{1}{2}\frac{d^2}{dx^2} + V(x)$, with the potential $V(x) = 0$ for $x > 0$ and $V(x) = \infty$ for $x \leq 0$. We have already evaluated this in Sec. 3.1 in Eq. (23). Putting $x = x_0 = \epsilon$ in Eq. (23) we get $Z_E = G(\epsilon, t \,|\epsilon, 0) = (1 - e^{-2\epsilon^2 t})/\sqrt{2\pi t}$. The path integral in the numerator in Eq. (41) is simply the propagator $\langle \epsilon | e^{-\hat{H} t} | \epsilon \rangle$ where the Hamiltonian $\hat{H} \equiv -\frac{1}{2}\frac{d^2}{dx^2} + pU(x)$ with a triangular potential $U(x) = x$ for $x > 0$ and $U(x) = \infty$ for $x \leq 0$. The Hamiltonian $\hat{H}$ has only bound states and discrete eigenvalues. Its eigenfunctions are simply shifted Airy functions and eigenvalues are given by the negative of the zeros of the Airy function. Expanding the propagator into its eigenbasis and finally taking the $\epsilon \to 0$ limit (for details see Ref. [49]), one derives the result

$$Q(0,t) = \int_0^\infty P(A,t)\, e^{-p\,A}\, dA = \sqrt{2\pi}\, (pt^{3/2}) \sum_{k=1}^{\infty} e^{-2^{-1/3}\, \alpha_k\, (pt^{3/2})^{2/3}} \tag{43}$$

where α_k's are the negative of the zeros of the Airy function. The result in Eq. (43) indicates that its inverse Laplace transform has the scaling form, $P(A,t) = t^{-3/2}\, f(At^{-3/2})$ where the Laplace transform of the scaling function $f(x)$ is given in Eq. (38).

Applications of the Airy Distribution Function: The Airy distribution function in Eq. (39) has appeared in a number of applications ranging from computer science and graph theory to physics. Below we mention some of these applications.

1. Cost function in data storage: One of the simplest algorithms for data storage in a linear table is called the linear probing with hashing (LPH) algorithm. It was originally introduced by D. Knuth [50] and has been the object of intense study in computer science due to its simplicity, efficiency and general applicability [44]. Recently it was shown [51] that the LPH algorithm gives rise to a correlated drop-push percolation model in one dimension that belongs to a different universality class compared to the ordinary site percolation model. Knuth, a pioneer in the analysis of algorithms, has indicated that this problem has had a strong influence on his scientific career [44]. The LPH algorithm is described as follows: Consider M items x_1, x_2, $\ldots$, x_M to be placed sequentially into a linear table with L cells labelled 1, 2, $\ldots$, L where $L \geq M$. Initially all cells are empty and each cell can contain at most one item. For each item x_i, a hash address $h_i \in \{1, 2, \ldots, L\}$ is assigned, i.e. the label h_i denotes the address of the cell to which x_i should go. Usually the hash address h_i is chosen randomly from the set $\{1, 2, \ldots, L\}$. The item x_i is inserted at its hash address h_i provided the cell labelled h_i is empty. If it is already occupied, one tries cells $h_i + 1$, $h_i + 2$, etc. until an empty cell is found (the locations of the cells are interpreted modulo L) where the item x_i is finally inserted. In the language of statistical physics, this is like a drop-push model. One starts with an empty periodic lattice. A site is chosen at random and one attempts to drop a particle there. If the target site is empty, the incoming particle occupies it and one starts the process with a new particle. If the target site is occupied, then the particle keeps hopping to the right until it finds an empty site which it then occupies and then one starts with a new particle and so on.

From the computer science point of view, the object of interest is the cost function $C(M, L)$ defined as the total number of unsuccessful probes encountered in inserting the M items into a table of size L. In particular, the total cost $C = C(L, L)$ in filling up the table is an important measure of the efficiency of the algorithm. The cost C is clearly a random variable, i.e. it has different values for different histories of filling up the table. A central question is: What is its pdf $P(C, L)$? It has been shown rigorously by combinatorial methods [44] that $P(C, L)$ has a scaling form for large L, $P(C, L) \simeq L^{-3/2} f(CL^{-3/2})$ where the scaling function $f(x)$ is precisely the Airy distribution function in Eq. (39) that describes the distribution of area under a Brownian excursion. To understand the connection between the two problems, consider any given history of the process where the table, starting initially with all sites empty, gets eventually filled up. We define

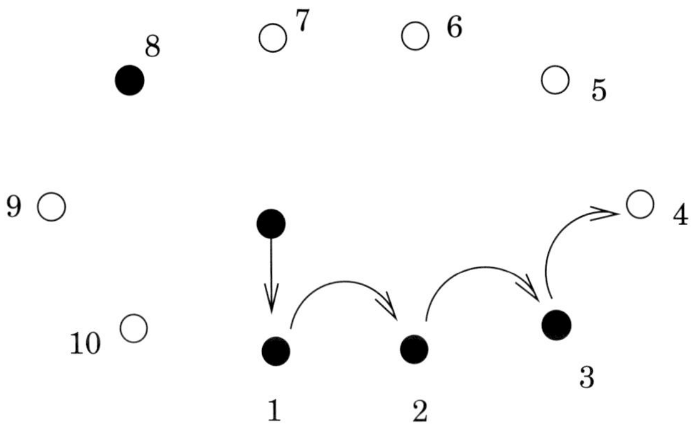

Fig. 2. The LPH algorithm for a table of 10 sites. The figure shows an incoming item which chose randomly the site 1 to drop, but since site 1 is already occupied, the incoming item keeps hopping to the right until it finds an empty cell at location 4 to which it gets absorbed.

a stochastic quantity X_i that measures the total number of attempts at site i till the end of the process in any given history. Clearly $X_i \geq 1$ and out of X_i attempts at site i, only one of the attempts (the first one) has been successful in filling up the site, the rest $(X_i - 1)$ of them had been unsuccessful. Thus, the total cost is $C = \sum_{i=1}^{L}(X_i - 1)$. Now, the site $(i-1)$ has been attempted X_{i-1} times, out of which only the first one was successful and the rest $(X_{i-1} - 1)$ attempts resulted in pushing the particle to the right neighbor i and thus each of these unsuccessful attempts at $(i-1)$ result in an attempt at site i. Thus, one can write a recursion relation

$$X_i = X_{i-1} - 1 + \xi_i \tag{44}$$

where ξ_i is a random variable that counts the number of direct attempts (not coming from site $(i-1)$) at site i. Thus Prob$(\xi = k)$ = Prob (the site i is chosen for direct hit k times out of a total L trials) $= \binom{L}{k}(1/L)^k(1 - 1/L)^{L-k}$, since for random hashing, the probability that site i is chosen, out of L sites, is simply $1/L$. Clearly the noise ξ has a mean value, $\langle \xi \rangle = 1$. If we now define $x_i = X_i - 1$, then x_i's satisfy

$$x_i = x_{i-1} + \eta_i \tag{45}$$

where $\eta_i = \xi_i - 1$ is a noise, independent from site to site, and for each site i, it is chosen from a binomial distribution. Note that $\langle \eta_i \rangle = \langle \xi_i \rangle - 1 = 0$. Thus, x_i's clearly represent a random walk in space from 0 to L with periodic boundary conditions. Moreover, since $X_i \geq 1$, we have $x_i \geq 0$,

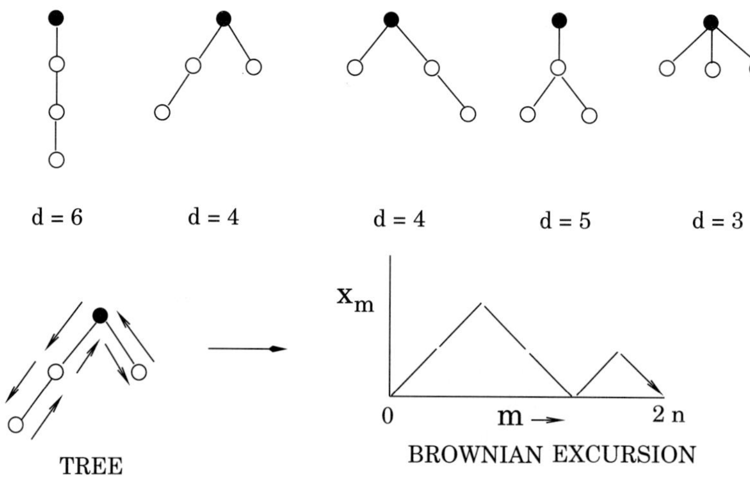

Fig. 3. The five possible rooted planar tree with (3 + 1) vertices. Each configuration has an associated total internal path length d listed below the configuration. Any given tree configuration, say the second one in the figure, has a one to one correspondence to a Dyck path, i.e. a configuration of a Brownian excursion (discrete time random walk).

indicating that it is a discrete version of a Brownian excursion and the total cost $C = \sum_{i=1}^{L}(X_i - 1) = \sum_{i=1}^{L} x_i$ is just the area under the Brownian excursion. For large number of steps L, the discrete and the continuum version share the same probability distribution, thus proving that the probability distribution of the total cost in LPH algorithm is precisely the same as that of the area under a Brownian excursion.

2. Internal path lengths on rooted planar trees: Rooted planar trees are important combinatorial objects in graph theory and computer science [52]. Examples of rooted planar trees with $n+1 = 4$ vertices are shown in Fig. 3. There are in general $C_{n+1} = \frac{1}{n+1}\binom{2n}{n}$ number of possible rooted planar tree configurations with $(n+1)$ vertices. For example, $C_1 = 1$, $C_2 = 1$, $C_3 = 2$, $C_4 = 5$, $C_6 = 14$, etc. — these are the Catalan numbers. An important quantity of interest is the total internal path length d of a tree which is simply the sum of the distances of all the n vertices from the root, $d = \sum_{i=1}^{n} d_i$, d_i being the distance of the ith vertex from the root. Each tree configuration has a particular value of d, e.g. in Fig. 3 the five different configurations have values $d = 6$, $d = 4$, $d = 4$, $d = 5$ and $d = 3$, respectively. Suppose that all C_{n+1} configurations of trees for a fixed n

are sampled with equal probability: what is the probability density $P(d, n)$ of the internal path length d? This problem again can be mapped [43] to the problem of the area under a Brownian excursion as shown in Fig. 3. Starting from the root of a planar tree with $(n+1)$ vertices, suppose one traverses the vertices of a tree as shown by the arrows in Fig. 3, ending at the root. We think of this route as the path of a random walker in one dimension. For each arrow pointing away from the root on the tree, we draw a step of the random walker with an upward slope. Similarly, for each arrow pointing to the root on the tree, we draw a step of the random walker with a downward slope. Since on the tree, one comes back to the root, it is evident by construction that the corresponding configuration of the random walker x_m is an excursion (i.e. it never goes to the negative side of the origin) that starts at the origin and ends up at the origin after $2n$ steps, $x_0 = 0$ and $x_{2n} = 0$. Such excursions of a discrete random walk are called Dyck paths. Now, the total internal path length d of any tree configuration is simply related to the total 'area' under a Dyck path via, $2d = \sum_{m=1}^{2n} x_m + n$, as can be easily verified. Now, for large n, Dyck paths essentially becomes Brownian excursions and the object $\sum_{m=1}^{2n} x_m$ is simply the area A_{2n} under a Brownian excursion over the time interval $[0, 2n]$. Since $A_{2n} \sim (2n)^{3/2}$ for large n, it follows that $d \simeq A_{2n}/2$. Therefore, its probability density $P(d, n)$ has a scaling form, $P(d, n) = \frac{1}{\sqrt{2}n^{3/2}} f(d/\sqrt{2}n^{3/2})$ where $f(x)$ is precisely the Airy distribution function in Eq. (39).

3. Maximal relative height distribution for fluctuating interfaces: Fluctuating interfaces have been widely studied over the last two decades as they appear in a variety of physical systems such as growing crystals, molecular beam epitaxy, fluctuating steps on metals and growing bacterial colonies [24]. The most well studied model of a fluctuating $(1+1)$-dimensional surfaces is the so-called Kardar–Parisi–Zhang (KPZ) equation [23] that describes the time evolution of the height $H(x, t)$ of an interface growing over a linear substrate of size L via the stochastic partial differential equation

$$\frac{\partial H(x,t)}{\partial t} = \frac{\partial^2 H(x,t)}{\partial x^2} + \lambda \left(\frac{\partial H(x,t)}{\partial x} \right)^2 + \eta(x,t)\,, \tag{46}$$

where $\eta(x, t)$ is a Gaussian white noise with zero mean and a correlator, $\langle \eta(x,t)\eta(x't') \rangle = 2\delta(x - x')\delta(t - t')$. If the parameter $\lambda = 0$, the equation becomes linear and is known as the Edwards–Wilkinson equation [22]. We consider the general case when $\lambda \geq 0$. The height is usually measured relative to the spatially averaged height, i.e. $h(x,t) = H(x,t) - \int_0^L H(x',t)dx'/L$. The joint probability distribution of the relative height field $P(\{h\},t)$ becomes time-independent as $t \to \infty$ in a finite system of size L. An important quantity that has created some interests recently [53–55] is the pdf of the maximal relative height (MRH) in the stationary state, i.e. $P(h_m, L)$ where

$$h_m = \lim_{t\to\infty} \max_x[\{h(x,t)\}, 0 \le x \le L]\,. \tag{47}$$

This is an important physical quantity that measures the extreme fluctuations of the interface heights. Note that in this system the height variables are strongly correlated in the stationary state. While the theory of extremes of a set of uncorrelated (or weakly correlated) random variables is well established [56], not much is known about the distribution of extremes of a set of strongly correlated random variables. Analytical results for such strongly correlated variables would thus be welcome from the general theoretical perspective and the system of fluctuating interfaces provides exactly the opportunity to study the extreme distribution analytically in a strongly correlated system. This problem of finding the MRH distribution was recently mapped [49, 54] again to the problem of the area under a Brownian excursion using the path integral method outlined in Sec. 3 and it was shown that for periodic boundary conditions, $P(h_m, L) = L^{-1/2} f(h_m/\sqrt{L})$ where $f(x)$ is again the Airy distribution function in Eq. (39). Interestingly, the distribution does not depend explicitly on λ. This is thus one of the rare examples where one can calculate analytically the distribution of the extreme of a set of strongly correlated random variables [49, 54].

4. Other applications: Apart from the three examples mentioned above, the Airy distribution function and its moments also appear in a number of other problems. For example, the generating function for the number of inversions in trees involves the Airy distribution function $f(x)$ [57]. Also, the moments M_n's of the function $f(x)$ appear in the enumeration of the connected components in a random graph [58]. Recently, it has been conjectured and subsequently tested numerically that the asymptotic pdf of the area of two-dimensional self-avoiding polygons is also given by the Airy distribution function $f(x)$ [59]. Besides, numerical evidence suggests that the area enclosed by the outer boundary of planar random loops is also distributed according to the Airy distribution function $f(x)$ [59].

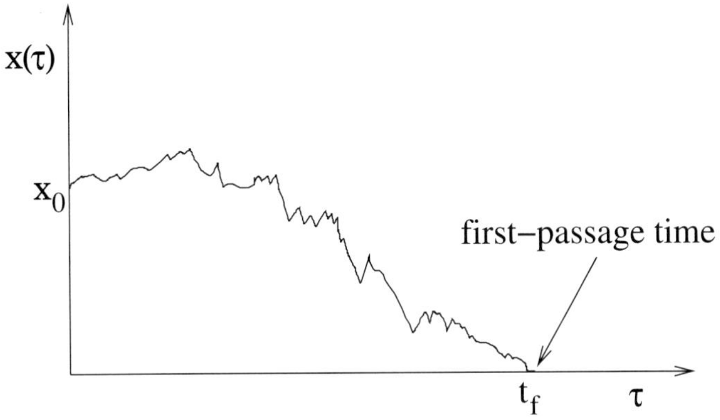

Fig. 4. A Brownian path $x(\tau)$, starting at x_0 at $\tau = 0$, crosses the origin for the first time at $\tau = t_f$, t_f being the first-passage time.

5. First-Passage Brownian Functional

So far we have studied the pdf of a Brownian functional over a fixed time interval $[0, t]$. In this section, we show how to compute the pdf of a Brownian functional over the time interval $[0, t_f]$ where t_f is the first-passage time of the process, i.e. t_f itself is random. More precisely, we consider a functional of the type

$$T = \int_0^{t_f} U(x(\tau))\, d\tau \tag{48}$$

where $x(\tau)$ is a Brownian path starting from $x_0 \geq 0$ at $\tau = 0$ and propagating up to time $\tau = t$ and $U(x)$, as before, is some specified function. The integral in Eq. (48) is up to the first-passage time t_f which itself is random in the sense that it varies from realization to realization of the Brownian path (see Fig. 4). Such functionals appear in many problems (some examples are given below) in physics, astronomy, queuing theory, etc. and we will generally refer to them as first-passage Brownian functionals.

We would like to compute the pdf $P(T|x_0)$ of T in Eq. (48) given that the Brownian path starts at the initial position x_0. As before, it is useful to consider the Laplace transform

$$Q(x_0) = \int_0^{\infty} e^{-pT} P(T|x_0) dT = \langle e^{-p\int_0^{t_f} U(x(\tau))d\tau} \rangle \tag{49}$$

where the r.h.s. is an average over all possible Brownian paths starting at x_0 at $\tau = 0$ and stopping at the first time they cross the origin. For brevity, we have suppressed the p dependence of $Q(x_0)$. Note that each path, starting from x_0, evolves via Eq. (10) where $\xi(t)$ is a delta correlated white noise. Note also that t_f varies from path to path. Thus at first sight, this seems to be a rather difficult problem to solve. However, as we will see now that in fact this problem is simpler than the previous problem over a fixed time interval $[0, t]$!

To proceed, we split a typical path over the interval $[0, t_f]$ into two parts: a left interval $[0, \Delta\tau]$ where the process proceeds from x_0 to $x_0 + \Delta x = x_0 + \xi(0)\Delta\tau$ in a small time $\Delta\tau$ and a right interval $[\Delta\tau, t_f]$ in which the process starts at $x_0 + \Delta x$ at time $\Delta\tau$ and reaches 0 at time t_f. The integral $\int_0^{t_f} U(x(\tau))\, d\tau$ is also split into two parts: $\int_0^{t_f} = \int_0^{\Delta\tau} + \int_{\Delta\tau}^{t_f}$. Since the initial value is x_0, one gets $\int_0^{\Delta\tau} U(x(\tau))\, d\tau = U(x_0)\Delta\tau$ for small $\Delta\tau$. Then, Eq. (49) can be written as

$$Q(x_0) = \langle e^{-p\int_0^{t_f} U(x(\tau))\, d\tau} \rangle = \langle e^{-pU(x_0)\Delta\tau} Q(x_0 + \Delta x) \rangle_{\Delta x}\,, \tag{50}$$

where we have used the fact that for the right interval $[\Delta\tau, t_f]$, the starting position is $x_0 + \Delta x = x_0 + \xi(0)\Delta\tau$, which itself is random. The average in the second line of Eq. (50) is over all possible realizations of Δx. We then substitute $\Delta x = \xi(0)\Delta\tau$ in Eq. (50), expand in powers of $\Delta\tau$ and average over the noise $\xi(0)$. We use the fact that the noise $\xi(t)$ is delta correlated, i.e. $\langle \xi^2(0) \rangle = 1/\Delta\tau$ as $\Delta\tau \to 0$. The leading order term on the right-hand side of Eq. (50) is independent of $\Delta\tau$ and is simply $Q(x_0)$ which cancels the same term on the left-hand side of Eq. (50). Collecting the rest of the terms we get

$$\left[\frac{1}{2}\frac{d^2Q}{dx_0^2} - pU(x_0)Q(x_0) \right] \Delta\tau + O((\Delta\tau)^2) = 0\,. \tag{51}$$

Equating the leading order term to zero provides us an ordinary differential equation

$$\frac{1}{2}\frac{d^2Q}{dx_0^2} - pU(x_0)Q(x_0) = 0 \tag{52}$$

which is valid in $x_0 \in [0, \infty]$ with the following boundary conditions: (i) When the initial position $x_0 \to 0$, the first-passage time t_f must also be 0. Hence the integral $\int_0^{t_f} U(x(\tau))\, d\tau = 0$. From the definition in Eq. (50), we get $Q(x_0 = 0) = 1$ and (ii) when the initial position $x_0 \to \infty$, the first-passage time $t_f \to \infty$, hence the integral $\int_0^{t_f} U(x(\tau))\, d\tau$ also diverges in this limit, at least when $U(x)$ is a nondecreasing function of x. The definition in Eq. (50) then gives the boundary condition, $Q(x_0 \to \infty) = 0$.

So, given a functional $U(x)$, the scheme would be to first solve the ordinary differential equation (52) with the appropriate boundary conditions mentioned above to obtain $Q(x_0)$ explicitly and then invert the Laplace transform in Eq. (49) to get the desired pdf $P(T|x_0)$ of the first-passage functional. As a simple test of this method, let us first consider the case $U(x) = 1$. In this case the functional $T = \int_0^{t_f} U(x(\tau))\, d\tau = t_f$ is the first-passage time itself. The differential equation (52) can be trivially solved and the solution satisfying the given boundary conditions is simply

$$Q(x_0) = e^{-\sqrt{2p}\, x_0}. \tag{53}$$

Inverting the Laplace transform with respect to p gives the pdf of the first-passage time

$$P(t_f|x_0) = \frac{x_0}{\sqrt{2\pi}} \frac{e^{-x_0^2/2t_f}}{{t_f}^{3/2}}, \tag{54}$$

which is identical to the result in Eq. (24) obtained by the path integral method. Below, we provide a few nontrivial examples and applications of this method.

5.1. *Area till the first-passage time*

Here we calculate the pdf of the area under a Brownian motion (starting at x_0) till its first-passage time [60]. Thus the relevant functional is $A = \int_0^{t_f} x(\tau) d\tau$ and hence $U(x) = x$. In Fig. 4, A is just the area under the curve over the time interval $[0, t_f]$. This problem has many applications in combinatorics and queuing theory. For example, an important object in combinatorics is the area of a lattice polygon in two dimensions [61]. A particular example of a lattice polygon is the rooted staircase polygon whose two arms can be thought of as two independent random walkers whose trajectories meet for the first time at the end of the polygon. The difference walk between these two arms then defines, in the continuum limit, a Brownian motion. The area of such a polygon can then be approximated, in the continuum limit, by the area under a single Brownian motion till its first-passage time [60]. This picture also relates this problem to the directed Abelian sandpile model [62] where t_f is just the avalanche duration and the area A is the size of an avalanche cluster. Another application arises in queueing theory, where the length of a queue l_n after n time steps evolves stochastically [61]. In the simplest approximation, one considers a random walk model, $l_n = l_{n-1} + \xi_n$ where ξ_n's are independent and identically distributed random variables which model the arrival and departure of new customers. When the two rates equal, $\langle \xi_n \rangle = 0$. In the large n limit, l_n can be approximated by a Brownian motion $x(\tau)$, whereupon t_f becomes the so-called 'busy' period (i.e. the time until the queue first becomes empty) and the area A then approximates the total number of customers served during the busy period.

Substituting $U(x) = x$ in Eq. (52), one can solve the differential equation with the prescribed boundary conditions and the solution is [60]

$$Q(x_0) = 3^{2/3}\Gamma(2/3)\mathrm{Ai}(2^{1/3}p^{1/3}x_0) \tag{55}$$

where $\mathrm{Ai}(z)$ is the Airy function. It turns out that this Laplace transform can be inverted to give an explicit expression for the pdf [60]

$$P(A|x_0) = \frac{2^{1/3}}{3^{2/3}\Gamma(1/3)} \frac{x_0}{A^{4/3}} \exp\left[-\frac{2x_0^3}{9A}\right]. \tag{56}$$

Thus the pdf has a power law tail for large $A \gg x_0^3$, $P(A|x_0) \sim A^{-4/3}$ and an essential singularity $P(A|x_0) \sim \exp[-2x_0^3/9A]$ for small $A \to 0$. Following the same techniques, one can also derive the pdf of the area till the first-passage time under a Brownian motion with a drift towards the origin — in this case the pdf has a stretched exponential tail for large A [60], $P(A|x_0) \sim A^{-3/4}\exp[-\sqrt{8\mu^3 A/3}]$ where μ is the drift.

Note the difference between the pdf of the area $P(A|x_0)$, under a Brownian motion till its first-passage time starting at x_0 at $\tau = 0$, as given in Eq. (56) and the pdf of the area under a Brownian excursion $P(A,t)$ in Eq. (43). In the latter case, the Brownian path is conditioned to start at $x_0 = 0$ at $\tau = 0$ and end at $x = 0$ at $\tau = t$ and one is interested in the statistics of the area under such a conditioned path over the *fixed* time interval t. In the former case on the other hand, one is interested in the area under a free Brownian motion starting at $x_0 > 0$ and propagating up to its first-passage time t_f that is *not fixed* but varies from one realization of the path to another.

5.2. *Time period of oscillation of an undamped particle in a random potential*

The study of transport properties in a system with quenched disorder is an important area of statistical physics [63]. The presence of a quenched disorder makes analytical calculations hard and very few exact results are known. Perhaps the simplest model that captures some complexities associated with the transport properties in disordered systems is that of a classical Newtonian particle moving in a one-dimensional random potential $\phi(x)$

$$m\frac{d^2x}{dt^2} + \Gamma\frac{dx}{dt} = F(x(t)) + \xi(t) \tag{57}$$

where $F(x) = -d\phi/dx$ is the force derived from the random potential $\phi(x)$, Γ is the friction coefficient and $\xi(t)$ is the thermal noise with zero mean and a delta correlator, $\langle\xi(t)\xi(t')\rangle = 2D\delta(t-t')$ with $D = k_B T/\Gamma$ by the Stokes–Einstein relation (3).

It turns out that even this simple problem is very hard to solve analytically for an arbitrary random potential $\phi(x)$. A special choice of the random potential where one can make some progress is the Sinai potential [27], where one assumes that $\phi(x) = \int_0^x \eta(x')dx'$. The variables $\eta(x)$'s have zero mean and are delta correlated $\langle\eta(x_1)\eta(x_2)\rangle = \delta(x_1 - x_2)$. Thus the potential $\phi(x)$ itself can be considered as a Brownian motion in space. In the overdamped limit when the frictional force is much larger than the inertial force, Eq. (57) then reduces to the Sinai model [27]

$$\Gamma\frac{dx}{dt} = F(x = x(t)) + \xi(t) \tag{58}$$

where the random force $F(x) = -d\phi/dx = \eta(x)$ is just a delta correlated white noise with zero mean: $\langle F(x)\rangle = 0$ and $\langle F(x)F(x')\rangle = \delta(x - x')$.

Here we consider a simple model [64] where the particle diffuses in the same Sinai potential $\phi(x) = \int_0^x \eta(x')dx'$, but we consider the opposite limit where the particle is *undamped*, i.e. $\Gamma = 0$ and is driven solely by the inertial force. For simplicity, we also consider the zero temperature limit where the thermal noise term drops out of Eq. (57) as well and one simply has

$$m\frac{d^2x}{dt^2} = F(x = x(t)) \tag{59}$$

where $F(x)$ is a same random Sinai force as mentioned above. We set $m = 1$ and assume that the particle starts at the origin $x = 0$ with initial velocity $v > 0$. Thus the particle will move to the right till it reaches a turning point x_c where the potential energy becomes equal to the kinetic energy, i.e. $\phi(x_c) = v^2/2$ and then it will move back to $x = 0$ with a velocity $-v$ (see Fig. 5). After returning to the origin with velocity $-v$, the particle will go to the left till it encounters the first turning point on the left of the origin where it will turn and then will return to the origin. Let T and T' denote the time for the particle to go from the origin to the turning

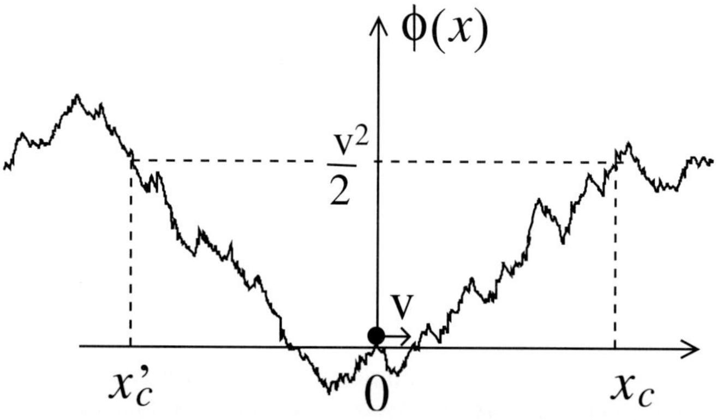

Fig. 5. A Newtonian particle in a Brownian potential with initial velocity v. The right/left turning points are shown as x_c and x'_c, respectively where the potential energy first becomes equal to the kinetic energy, $\phi = v^2/2$.

point at the right and to the one at the left, respectively. Thus the particle will oscillate between the two turning points nearest to the origin on either side and the time period of oscillation is $T_{\rm osc} = 2(T + T')$. Note that the variables T and T' will vary from one sample of quenched disorder to another. The goal is to compute the probability distribution of T and T' and hence that of $T_{\rm osc}$. Since $\phi(x)$ is a Brownian motion in x, it follows from its Markov property that $\phi(x)$ for $x > 0$ and for $x < 0$ are completely independent of each other. Thus T and T' are also independent and by symmetry, have identical distributions. The distribution of $T_{\rm osc}$ can then be easily calculated by convolution.

To compute the pdf $P(T)$ of T (starting at $x_0 = 0$), we first express T as a functional of the Brownian potential

$$T = \int_0^{x_c} \frac{dx}{\sqrt{v^2 - 2\phi(x)}} \tag{60}$$

where x_c is defined as the point where $\phi(x_c) = v^2$. On identifying the space as the time $x \equiv \tau$ and the random potential ϕ as the trajectory of a random walk in space x, i.e. $\phi \leftrightarrow x$, $x \leftrightarrow \tau$, T in Eq. (60) is of the general form in Eq. (48) with $U(x) = 1/\sqrt{v^2 - 2x}$ and $x_c = t_f$ denoting the first-passage time to the level $x = v^2/2$, starting at x_0. Following the general scheme, we need to solve the differential equation (52), now valid for $-\infty \le x_0 \le v^2/2$, with $U(x) = 1/\sqrt{v^2 - 2x}$ and the boundary conditions, $Q(x_0 \to -\infty) = 0$ and $Q(x_0 \to v^2/2) = 1$. Upon finding the solution one needs to put $x_0 = 0$

and then invert the Laplace transform. This can be done explicitly and one obtains [64]

$$P(T) = \frac{2^{2/3}v^2}{3^{4/3}\Gamma(2/3)}\frac{1}{T^{5/3}}\exp\left[-\frac{2v^3}{9T}\right]. \tag{61}$$

This is one of the rare examples of an exact result on a transport property in a quenched disorderd system, thus illustrating the power of the approach outlined in this section.

5.3. *Distribution of the lifetime of a comet in solar system*

In this final subsection we provide an example from astrophysics [65] where the general technique of the first-passage Brownian functional is applicable. A comet enters a solar system with a negative energy $E_0 < 0$ and keeps orbiting around the sun in an elliptical orbit whose semimajor axis length a is determined by the relation $E_0 = -GM/2a$ where G is the gravitational constant and M is the mass of the sun. It was first pointed out by Lyttleton [66] that the energy of the comet gets perturbed by Jupiter each time the comet visits the neighborhood of the sun and the planets and successive perturbations lead to a positive energy of the comet which then leaves the solar system. It is convenient to work with the negative energy $x = -E > 0$ of the comet. We assume that the comet enters the solar system with initial negative energy x_0 and has values of x equal to $x_1, x_2, \ldots, x_{t_f}$ at successive orbits till the last one labelled by t_f when its value of x crosses 0 (energy becomes positive) and it leaves the solar system. The lifetime of the comet is given by

$$T = U(x_0) + U(x_1) + \ldots U(x_{t_f}) \tag{62}$$

where $U(x)$ is the time taken to complete an orbit with negative energy $x > 0$. According to Kepler's third law, $U(x) = cx^{-3/2}$ where c is an constant which we set to $c = 1$ for convenience. Moreover, a simple way to describe the perturbation due to Jupiter is by a random walk model, $x_n = x_{n-1} + \xi_n$ where ξ_n is the noise induced by Jupiter and is assumed to be independent from orbit to orbit [65]. Within this random walk theory, the lifetime of a comet in Eq. (62), in the continuum limit becomes a first-passage Brownian functional [65]

$$T = \int_0^{t_f} [x(\tau)]^{-3/2} d\tau \tag{63}$$

where the random walk starts at x_0 and ends at its first-passage time t_f when it first crosses the origin. The pdf $P(T|x_0)$ was first obtained by Hammersley [65]. Here we show how to obtain this result using the general approach outlined here for first-passage Brownian functionals.

Following our general scheme, we thus have $U(x) = x^{-3/2}$ in the differential Eq. (52). The solution, satisfying the proper boundary conditions, can be easily found

$$Q(x_0) = 16p{x_0}^{1/2}K_2(\sqrt{32p}{x_0}^{1/4}) \tag{64}$$

where $K_2(z)$ is the modified Bessel function of degree 2. Next, we need to invert the Laplace transform in Eq. (64) with respect to p. This can be done by using the following identity

$$\int_0^\infty y^{-\nu-1}e^{-py-\beta/y}dy = 2\left(\frac{p}{\beta}\right)^{\nu/2} K_\nu(2\sqrt{\beta p}). \tag{65}$$

Choosing $\beta = 8\sqrt{x_0}$, we can invert the Laplace transform to obtain the exact pdf $P(T|x_0)$ of the lifetime of a comet

$$P(T|x_0) = \frac{64x_0}{T^3}\exp\left[-\frac{8\sqrt{x_0}}{T}\right]. \tag{66}$$

It is worth pointing out that in all three examples above, the pdf $P(T|x_0)$ of the first-passage Brownian functional has a power law tail $P(T|x_0) \sim T^{-\gamma}$ for large T and an essential singularity in the limit $T \to 0$. While the exponent of the power law tail can be easily obtained using a scaling argument, the essential singular behavior at small T is not easy to obtain just by a scaling argument.

6. Conclusion

In this article I have provided a brief and pedagogical review of the techniques to calculate the statistical properties of functionals of one-dimensional Brownian motion. It also contains a section devoted to 'first-passage' Brownian functional, a quantity that appears in many problems but the techniques to calculate its properties are somewhat less known compared to the standard Feynman–Kac formalism for the usual Brownian functional. A simple backward Fokker–Planck approach is provided here to calculate the probability distribution of a first-passage Brownian functional. Several examples and applications of the standard Brownian functionals as well as the first-passage Brownian functionals from physics, probability theory, astronomy and in particular, from computer science are provided.

The techniques detailed in this article are valid for free Brownian motion in one dimension. However, they can be easily generalized to study the functionals of a Brownian motion in an external potential. The external potential can represent e.g. a constant drift [28, 29, 60] or a harmonic potential [30]. Alternately, the external potential can be random as in a disordered system. The backward Fokker–Planck approach reviewed here has been particularly useful in calculating exactly the disorder averaged distributions of Brownian functionals in the Sinai model [28, 39, 67].

There are several open directions for future research. For example, to the best of my knowledge, the properties of first-passage Brownian functionals have so far not been studied in disordered systems. The techniques discussed here could be useful in that direction. Though there have been few studies of Brownian functionals in higher dimensions, there are still many open problems with direct relation to experiments [12] and more studies in that direction would be welcome. Finally, the discussion in this article is tilted to the simple Brownian motion which is a Gaussian as well as a Markov process. In many real systems, the relevant stochastic process often is non-Gaussian and/or non-Markovian. It would certainly be interesting to study the properties of functionals of such stochastic processes.

In summary, I hope I have been able to convey to the reader the beauty and the interests underlying Brownian 'functionalogy' with its diverse applications ranging from physics and astronomy to computer science, making it a true legacy of Albert Einstein whose 1905 paper laid the basic foundation of this interesting subject.

Acknowledgments

It is a pleasure to thank my collaborators A. J. Bray, A. Comtet, C. Dasgupta, D. S. Dean, A. Dhar, M. J. Kearney, and S. Sabhapandit with whom I have worked on Brownian functionals in the recent past. I also acknowledge useful discussions on Brownian motion and related topics with M. Barma, J. Desbois, D. Dhar, P. L. Krapivsky, S. Redner, C. Sire, C. Texier, M. Yor and R. M. Ziff.

References

[1] Einstein, A. [1905] On the movement of small particles suspended in stationary liquids required by the molecular-kinetic theory of heat, *Ann. der Physik* **17**, 549–560.

[2] Einstein, A. [1905] On the electrodynamics of moving bodies, *Ann. der Physik* **17**, 891–921.

[3] Einstein, A. [1905] An heuristic point of view about the production and transformation of light, *Ann. der Physik* **17**, 132–148.

[4] von Smoluchowski, M. [1906] Zur kinetischen theorie der Brownschen molekularbewegung und der suspensionen, *Ann. der Physik* **21**, 756–780.

[5] Langevin, P. [1908] *Comptes Rendues* **146**, 530–533.

[6] Frey, E. and Kroy, K. [2005] Brownian motion: a paradigm of soft matter and biological physics, *Ann. der Physik* **14**, 20–50.

[7] Duplantier, B. [2005] Le mouvement Brownien, "divers et ondoyant", *Séminaire Poincaré* **1**, 155–212.

[8] Sokolov, I. M. and Klafter, J. [2004] From diffusion to anamalous diffusion: a century after Einstein's Brownian motion, cond-mat/0411032. Hänggi P. and Marchesoni, F. [2005] Introduction: 100 years of Brownian motion, *Chaos* **15**, 026101–026105; Chowdhury, D. [2005] 100 years of Einstein's theory of Brownian motion: from pollen grains to protein trains, cond-mat/0504610.

[9] Kac, M. [1949] On distribution of certain Wiener functionals, *Trans. Am. Math. Soc.* **65**, 1–13; [1951] On some connections between probability theory and differential and integral equations, *Proc. Second Berkeley Symp. Mathematical Statistics and Probability, 1950* (University of California Press, Berkeley and Los Angeles), 189–215.

[10] Yor, M. [1992] Some aspects of Brownian motion, *Lectures in Mathematics* (ETH Zurich, Birkhausser).

[11] Dufresne, D. [1990] The distribution of perpetuity, with applications to risk theory and pension funding, *Scand. Act. J.* 39–79; Geman, H. and Yor, M. [1993] Bessel processes, Asian options and perpetuities, *Math. Fin.* **3**, 349–375.

[12] For a recent review see Comtet, A., Desbois, J. and Texier, C., Functionals of the Brownian motion, localization and metric graphs, cond-mat/0504513.

[13] Feynman, R. P. and Hibbs, A. R. [1965] *Quantum Mechanics and Path Integrals* (McGraw-Hill, New York).

[14] Majumdar, S. N. [1999] Persistence in nonequilibrium systems, *Current Sci.* **77**, 370–375.

[15] Majumdar, S. N., Sire, C., Bray, A. J. and Cornell, S. J. [1996] Nontrivial exponent for simple diffusion, *Phys. Rev. Lett.* **77**, 2867–2870; Derrida, B., Hakim, V. and Zeitak, R. [1996] Persistent spins in the linear diffusion approximation of phase ordering and zeros of stationary Gaussian processes, *Phys. Rev. Lett.* **77**, 2871–2874.

[16] Wong, G. P., Mair, R. W., Walsworth, R. L. and Cory, D. G. [2001] Measurement of persistence in 1D diffusion, *Phys. Rev. Lett.* **86**, 4156–4159.
[17] Watson, A. [1996] Persistence pays off in defining history of diffusion, *Science* **274**, 919.
[18] Chandrasekhar, S. [1943] Stochastic problems in physics and astronomy, *Rev. Mod. Phys.* **15**, 1–89.
[19] Feller, W. [1968] *An Introduction to Probability Theory and Its Applications*, Vols. I and II (Wiley, New York).
[20] Redner, S. [2001] *A Guide to First-passage Processes* (Cambridge University Press, Cambridge).
[21] Lévy, P. [1939] Sur certains processus stochastiques homogènes, *Compositio Mathematica*, **7**, 283–339.
[22] Edwards, S. F. and Wilkinson, D. R. [1982] The surface statistics of a granular agregate, *Proc. R. Soc. London A* **381**, 17–31.
[23] Kardar, M., Parisi, G. and Zhang, Y.-C. [1986] Dynamical scaling of growing interfaces, *Phys. Rev. Lett.* **56**, 889–892.
[24] Barabasi, A. L. and Stanley, H. E. [1995] *Fractal Concepts in Surface Growth* (Cambridge University Press, Cambridge, England); Krug, J. [1997] Origins of scale invariance in growth processes, *Adv. Phys.* **46**, 139–282. Halpin-Healy, T. and Zhang, Y.-C. [1995] Kinetic roughening phenomena, stochastic growth, directed polymers and all that, *Phys. Rep.* **254**, 215–414.
[25] Foltin, G., Oerding, K., Racz, Z., Workman, R. L. and Zia, R. K. P. [1994] Width distribution for random-walk interfaces, *Phys. Rev. E* **50**, R639–R642; Racz, Z. and Plischke, M. [1994] Width distribution for 2+1 dimensional growth and deposition processes, *Phys. Rev. E* **50**, 3530–3537.
[26] Yor, M. [2000] Exponential functionals of Brownian motion and related processes, (Springer, Berlin).
[27] Sinai, Ya. G. [1982] The limiting behavior of a one-dimensional random walk in a random medium, *Teor. Veroyatnost. i Primenen.* **27**, 247–258.
[28] Comtet, A., Monthus, C. and Yor, M. [1998] Exponential functionals of Brownian motion and disordered systems, *J. Appl. Probab.* **35**, 255–271.
[29] Comtet, A. and Texier, C. [1999] Universality of the Wigner time delay distribution for one-dimensional random potentials, *Phys. Rev. Lett.* **82**, 4220–4223.
[30] Majumdar, S. N. and Bray, A. J. [2002] Large deviation functions for nonlinear functionals of a Gaussian stationary Markov process, *Phys. Rev. E* **65**, 051112.
[31] Cifarelli, D. M. and Regazzini, E. [1975] Contributi intorno ad un test per l'homogeneita tra du campioni, *Giornale Degli Economiste*, **34**, 233–249.
[32] Shepp, L. A. [1982] On the integral of the absolute value of the pinned Wiener process, *Ann. Prob.* **10**, 234–239.
[33] Altshuler, B. L., Aronov, A. G. and Khmelnitsky, D. E. [1982] Effects of electron-electron collisions with small energy transfers on quantum localization, *J. Phys. C: Solid. St. Phys.* **15**, 7367–7386.

[34] Lamperti, J. [1958] An occupation time theorem for a class of stochastic processes, *Trans. Am. Math. Soc.* **88**, 380–387.
[35] Dornic, I. and Godrèche, C. [1998] Large deviations and nontrivial exponents in coarsening systems, *J. Phys. A.: Math. Gen.* **31**, 5413–5429; Newman, T. J. and Toroczkai, Z. [1998] Diffusive persistence and "sign-time" distribution, *Phys. Rev. E* **58**, R2685–R2688.
[36] Dhar, A. and Majumdar, S. N. [1999] Residence time distribution for a class of Gaussian markov processes, *Phys. Rev. E* **59**, 6413–6418; De Smedt, G., Godreche, C. and Luck, J. M. [2001] Statistics of occupation time for a class of Gaussian markov processes, *J. Phys. A: Math. Gen.* **34**, 1247–1269.
[37] Godrèche, C. and Luck, J. M. [2001] Statistics of occupation time of renewal processes, *J. Stat. Phys.* **104**, 489–524.
[38] Majumdar, S. N. and Dean, D. S. [2002] Exact occupation time distribution in a non-Markovian sequence and its relation to spin glass models, *Phys. Rev. E* **66**, 041102.
[39] Majumdar, S. N. and Comtet, A. [2002] Local and occupation time of a particle diffusing in a random medium, *Phys. Rev. Lett.* **89**, 060601.
[40] Bel, G. and Barkai, E. [2005] Weak ergodicity breaking in the continuous-time random walk, *Phys. Rev. Lett.* **94**, 240602.
[41] Darling, D. A. [1983] On the supremum of certain Gaussian processes, *Ann. Prob.* **11**, 803–806.
[42] Louchard, G. [1984] Kac's formula, Levy's local time and Brownian excursion, *J. Appl. Prob.* **21**, 479–499.
[43] Takacs, L. [1991] A Bernoulli excursion and its various applications, *Adv. Appl. Prob.* **23**, 557–585; [1995] Limit distributions for the Bernoulli meander, *J. Appl. Prob.* **32**, 375–395.
[44] Flajolet, P., Poblete, P. and Viola, A. [1998] On the analysis of linear probing hashing, *Algorithmica.* **22**, 490–515.
[45] Flajolet, P. and Louchard, G. [2001] Analytic variations on the airy distribution, *Algorithmica.* **31**, 361–377.
[46] Perman, M. and Wellner, J. A. [1996] On the distribution of Brownian areas, *Ann. Appl. Prob.* **6**, 1091–1111.
[47] Abramowitz, M. and Stegun, I. A. [1973] *Handbook of Mathematical Functions* (Dover, New York).
[48] Csörgö, M., Shi, Z. and Yor, M. [1999] Some asymptotic properties of the local time of the uniform empirical processes, *Bernoulli.* **5**, 1035–1058.
[49] Majumdar, S. N. and Comtet, A. [2005] Airy distribution function: from the area under a Brownian excursion to the maximal height of fluctuating interfaces, *J. Stat. Phys.* **119**, 777–826.
[50] Knuth, D. E. [1998] The art of computer programming, Vol. 3, *Sorting and Searching* (Addison-Wesley, Reading, MA).
[51] Majumdar, S. N. and Dean, D. S. [2002] Exact solution of a drop-push model for percolation, *Phys. Rev. Lett.* **89**, 115701.
[52] Harary, F. [1988] *Graph Theory* (Addison-Wesley, Reading, MA).
[53] Raychaudhuri, S., Cranston, M., Przybyla, C. and Shapir, Y. [2001] Maximal height scaling of kinetically growing surfaces, *Phys. Rev. Lett.* **87**, 136101.

[54] Majumdar, S. N. and Comtet, A. [2004] Exact maximal height distribution of fluctuating interfaces, *Phys. Rev. Lett.* **92**, 225501.
[55] Guclu, H. and Korniss, G. [2004] Extreme fluctuations in small-world networks with relaxional dynamics, *Phys. Rev. E* **69**, 065104(R).
[56] Gumbel, E. J. [1958] *Statistics of Extremes* (Columbia University Press, New York).
[57] Mallows, C. L. and Riordan, J. [1968] The inversion enumerator for labelled trees, *Bull. Am. Math. Soc.* **74**, 92–94; Gessel, I., Sagan, B. E. and Yeh, Y.-N. [1995] Enumeration of trees by inversion, *J. Graph Th.* **19**, 435–459.
[58] Wright, E. M. [1977] The number of connected sparsely edged graphs, *J. Graph Th.* **1**, 317–330.
[59] Richard, C., Guttmann, A. J. and Jensen, I. [2001] Scaling function and universal amplitude combinations for self avoiding polygons, *J. Phys. A: Math. Gen.* **34**, L495–L501; Richard, C. [2002] Scaling behaviour of the two-dimensional polygons models, *J. Stat. Phys.* **108**, 459–493; [2004] Area distribution of the planar random loop boundary, *J. Phys. A: Math. Gen.* **37**, 4493–4500.
[60] Kearney, M. J. and Majumdar, S. N. [2005] On the area under a continuous time Brownian motion till its first-passage time, *J. Phys. A: Math. Gen.* **38**, 4097–4104.
[61] See Kearney, M. J. [2004] On a random area variable arising in discrete-time queues and compact directed percolation, *J. Phys. A: Math. Gen.* **37**, 8421–8431 and references therein.
[62] Dhar, D. and Ramaswamy, R. [1989] Exactly solved model of self-organized critical phenomena, *Phys. Rev. Lett.* **63**, 1659–1662.
[63] Bouchaud, J. P. and Georges, A. [1990] Anomalous diffusion in disordered media: statistical mechanics, models and physical applications, *Phys. Rep.* **195**, 127–293.
[64] Dean, D. S. and Majumdar, S. N. [2001] The exact distribution of the oscillation period in the underdamped one-dimensional Sinai model, *J. Phys. A: Math. Gen.* **34**, L697–L702.
[65] Hammersley, J. M. [1961] On the statistical loss of long period comets from the solar system II, *Proc. Fourth Berkeley Symposium on Mathematical Statistics and Probability*, Vol. 3 (University of California Press, Berkeley and Los Angeles), 17–78.
[66] Lyttleton, R. A. [1953] *The Comets and Their Origin* (Cambridge University Press, Cambridge).
[67] Majumdar, S. N. and Comtet, A. [2002] Exact asymptotic results for persistence in the Sinai model with arbitrary drift, *Phys. Rev. E* **66**, 061105.

CHAPTER 7

Bose–Einstein Condensation: Where Many Become One and So There is Plenty of Room at the Bottom

****

N. KUMAR

Raman Research Institute, Bangalore 560 080, India

Classically identical particles become quantum mechanically indistinguishable. Satyendra Nath Bose taught us, in 1924, how to correctly count the distinct microstates for the indistinguishables, and for a gas of light quanta (later photons), whose number is *not conserved*, e.g. can vary with temperature, he gave a proper derivation of Planck's law of black body radiation. Einstein, in 1925, generalized the Bose statistics to a quantum gas of material particles whose *number is now fixed*, or conserved, e.g. ^{4}He, and thus opened a new direction in condensed matter physics: He showed that for low enough temperatures ($\sim$1 Kelvin and below), a macroscopic number of the particles must accumulate in the lowest one-particle state. This degenerate gas with an extensively occupied single one-particle state is the Bose–Einstein condensate, now called BEC. (Fragmented BEC involving a multiplicity of internal states of non-scalar Bose atoms is, however, also realizable now.) Initially thought to be a pathology of an ideal non-interacting Bose system, the BEC turned out to be robust against interactions. Thus, the Bose–Einstein condensation is a quantum phase transition, but one with a difference — it is a purely quantum statistical effect, and requires no inter-particle interaction for its occurrence. Indeed, it happens in spite of it. The condensate fraction, however, diminishes with increasing interaction strength — to less than ten per cent for ^{4}He. The BEC turned out to underlie superfluidity, namely that the superfluid may flow through finest atomic capillaries without any viscosity. Interaction, however, seems essential to superfluidity. But, the precise connection between BEC and the superfluidity remains elusive. Thus, for example, we may have superfluidity in two-dimensions where there is no condensate! Seventy years later now, the BEC has come alive with the breakthrough in 1995 when near-ideal BEC was created in dilute alkali gases of ^{87}Rb and ^{23}Na atoms *cooled in the gaseous state* down to nanokelvins and localized in a trap. There are reasons why we ought to be mindful of the BEC — if only because here even the interaction between the particles is tunable at will — the sign as well as the strength of it. BEC has now become an ideal laboratory for basic and condensed matter experiments, and for high resolution applications. Properly viewed, it is indeed a new state of matter. This article is about the saga of BEC that really began with Einstein in 1925.

Let me begin with what may seem like an apology, which it is certainly not, and a touch of history of Bose–Einstein condensation, that it is. The point is that if we go strictly by the calendar, then the Bose–Einstein condensation does not belong in the miracle year of 1905, which is being observed as the World Year of Physics by many learned bodies around the world. In fact, Bose–Einstein condensation came in twenty years too late, in the year 1925 when Einstein, already famous at 45, derived the condensate for a degenerate quantum gas of permanent particles, i.e. of fixed number, such as helium (^{4}He), as a necessary consequence following from the novel quantum statistics [1] proposed a year earlier in 1924 by the young Indian lecturer, Satyendra Nath Bose at 25, then at the Dacca University, for the gas of light quanta (later photons). Bose had shown how to correctly count the distinct distributions (microstates or complexions) of indistinguishable objects (particles) among the distinguishable boxes (phase-space cells) with no restrictions on the occupation numbers. Actually, Bose had proposed this new quantum statistics in an attempt to give a logical derivation of the celebrated Planck law of black body radiation (Fig. 1), without the *ad hoc* assumptions that Planck had to make. His paper was, however, turned down by the editors of the *Philosophical Magazine*. Convinced that he was right, Bose sent his manuscript to Einstein. It was Einstein who first saw the important conceptual advance – the indistinguishability of identical particles — in Bose's work. At his (Bose's) request, contained in the letter accompanying the manuscript, Einstein personally translated it in German and got it published speedily in *Zeitschrift fuer Physik*, the prestigious physics journal of the time. (It is of certain historical note that in his letter young Bose had addressed Einstein as 'Respected master' in the typically Indian tradition of Ekalavya. They met only later in 1925 in Berlin. Einstein also promoted Bose's work in the Prussian Academy of Sciences.) Einstein at once saw the deep connection between Bose's view of the Planck thermal radiation as a gas of massless light quanta and an ideal quantum gas of identical material particles of non-zero mass, such as helium or hydrogen. The crucial point was that of indistinguishability of the identical quantum particles and the quantum-statistically correct way of counting them. As an act of pure transference, Einstein applied [2–5] the quantum statistics of Bose to a gas of identical particles, but with the

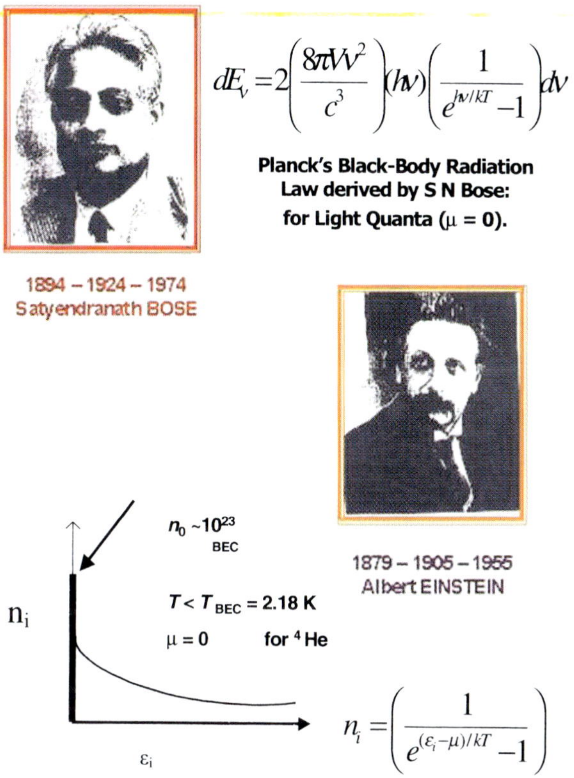

Fig. 1. Bose–Einstein condensation and quantum statistical phase transition: From the photon gas to a gas of indistinguishable molecules.

proviso that, unlike the case of photons, the number of particles must be conserved now. Hence, the introduction of a chemical potential that, at low enough temperatures, led to the condensation [4] (that now bears his name jointly with the name of Bose, see Fig. 1). This condensation was much to the disbelief of many leading physicists of the time; among them were Planck, Schrdinger, and Einstein himself perhaps. Einstein never returned to Bose–Einstein condensate (BEC) after 1925.

But, all this was about twenty years after the miracle year of 1905. The question then is why include BEC in this centennial issue. One obvious answer is that BEC was a landmark in the developments that followed the quantum revolution unfolding at the turn of the 20th century. C. N. Yang has remarked to the effect that just BEC alone would have assured a place for Einstein, and certainly for Bose, in the annals of physics. There is, however, to my mind, yet another, more efficient cause for its inclusion here. While it is true that BEC was not born in the miracle year of 1905, it did have a miraculous resurrection seventy years later in 1995, nearer our own times, when the near-ideal BEC was realized experimentally in a dilute gas of alkali atoms at a few billionths of a degree Kelvin (nanokelvins) above the absolute zero of temperature! For a review, see Refs. 6–8. This was an ideal condensation (one hundred per cent!) — a dream for a quantum condensed matter physicist! Indeed, the term BEC came into being just then — for this new state of matter. The saga of BEC that really began in 1925 with Einstein, lived through 1995 to our own times, and will certainly live on far beyond. See the timeline at the end. Thus justified, let us turn now to the physics of BEC.

Bose–Einstein condensation: The phenomenon

Normally, a gas such as the air in a room has its molecules distributed over a broad range of energy-momentum. This is familiar from the Maxwell distribution of velocities known from the 19th century. The velocity distribution is smooth and sparse — the occupancy number of an elementary molecular phase space cell ($\boldsymbol{dpdq} \sim h^3$) being typically <1. As the temperature is lowered, however, the situation begins to change. The gas condenses into liquid and then into the solid state, through successive phase transitions at which some of the thermodynamic quantities become singular. All these transitions involve interactions among the particles, e.g. the long-range van der Waals attraction. Something much more subtle happens for certain gases, such as helium that remains fluid down to the absolute zero of temperature under its own vapor pressure. (Helium (^{4}He) can be solidified only under pressure of about 25 atmospheres. This is because of its low atomic mass and weak inter-atomic attraction, giving it a large zero point energy — we call it a quantum liquid.) The helium (^{4}He), however, does undergo a phase transition at a critical temperature $T_c = 2.18$ K — a second order phase transition at which its specific heat at constant volume has a logarithmic singularity (λ-shaped and hence the name λ-point for the transition). A non-interacting (ideal) Bose gas has a gentler singularity, namely a cusp — a third order transition in the sense of Ehrenfest. The interaction drives it to the second order. Very spectacularly, the lower-temperature phase (called He II) turns out to be superfluid (with zero viscosity) while the higher-temperature phase (He I) remains normal. At a deeper level, however, for $T < T_c$, the velocity distribution ceases to be sparse, and a finite fraction of the ^{4}He atoms accumulates in the zero-momentum state. *This macroscopic, extensive occupation of the single one-particle state is the Bose–Einstein condensation*, or BEC for short. It is driven not by any inter-particle attraction, but is a purely quantum statistical effect. Much of this was, of course, not known in 1925. Einstein was led to the condensation from his close examination of the ideas that followed from the new quantum statistics that Bose had proposed in the previous year, 1924, for the gas of light quanta in the black body radiation. Einstein clearly saw the deep concept of indistinguishability of identical particles that was implicit in Bose's derivation of the Planck radiation law. All he really had to do then was to replace the gas of massless photons (with the relativistic dispersion relation $p = h\nu/c$) by the gas of material particles (with non-zero mass and the non-relativistic dispersion relation $E = p^2/2m$), and to introduce a chemical potential μ to ensure a fixed particle number N. A brief account of essentially his derivation of the condensate, and the underlying quantum gas statistics, is given below.

BEC and the quantum gas statistics

All statistics is about counting. And the quantum gas statistics is about counting the indistinguishables. It is concerned with finding the most probable distribution of the gas molecules over the molecular phase space subject to certain given constraints or subsidiary conditions, e.g. given total energy and the total number of particles. The term probability is used here in the sense of Planck, namely, that the probability of a macrostate (coarse grained macroscopic description) is determined by (proportional to) the number of distinct microstates (fine-grained microscopic descriptions or complexions) that are consistent with it (assuming that these microstates are degenerate in energy and equally probable). This is a problem in combinatorics — in how many ways (W) can we distribute N objects (particles) among Z boxes (phase-space cells) with the i^h box containing n_i objects. For the classical case (classical gas statistics) the boxes are, of course, distinct, but the N objects, though identical, are also distinguishable, and we have the classical Boltzmann result, $W_c = N!/\Pi_i n_i!$ Here, permuting (exchanging) the n_i objects within the ith box obviously creates no new microstates; but a permutation involving exchange of the particle between different boxes does create new microstates, and must be counted as such. Now, let the classically identical objects become quantum mechanically indistinguishable, as it must be in the case of quantum gas statistics. Then the permutations of the objects even between different boxes must be discounted — indeed, a permutation of the indistinguishables generates no new microstates! Then, the Boltzmannian number of microstates (complexions) for the classical case, W_c ($\gg 1$), must be replaced in the quantum case by the correctly counted $W_Q = 1$. This is the essence of indistinguishability and of the quantum statistics that Einstein had made use of.

BEC derived

Proceeding with Einstein, consider an ideal (non-interacting) gas of indistinguishable molecules for which the molecular phase-space lying in the energy shell $E_v \pm 1/2\Delta E_v$ has the number of elemental phase–space cells Z_ν given by

$$Z_\nu = (2\pi V/h^3)(2m)^{3/2}E_v^{1/2}dE_v\,, \tag{1}$$

(h being the volume of the elementary phase space cell after Planck).

We can now distribute N_v of the indistinguishable molecules among the Z_ν distinct cells in W_v ways, where

$$W_v = (N_v + Z_\nu - 1)!/N_v!(Z_\nu - 1)! \tag{2}$$

(To see that this is so, just imagine placing Z_ν partitions separating the N_v objects arranged on a line, and count the number of ways of doing this. This is essentially same as Bose's way of defining a microstate in terms of the set of occupation numbers of the cells; or equivalently, distributing the distinguishable (Z_ν) cells among the indistinguishable (N_v) molecules.) It is assumed here that $Z_\nu \gg 1$, which is true for a gas extended over a large volume with a phase–space that is a continuum. The total number of microstates for the N indistinguishable molecules distributed over the total phase–space is then

$$W = \Pi W_v\,. \tag{3}$$

The rest follows the standard exercise in maximizing the associated entropy (S) function subject to the subsidiary conditions, or constraints, of the given number (N) and energy (E):

$$\begin{aligned} S &= k_B \propto nW\,,\\ N &= \Sigma N_v\,, \\ E &= \Sigma E_v N_v\,. \end{aligned} \tag{4}$$

The constraints are to be imposed through the introduction of the corresponding Lagrange multipliers. Maximization of the entropy function is facilitated in the thermodynamic limit ($N \to \infty$, $V \to \infty$, with N/V = number density; n = constant) through the Stirling approximation for the otherwise tyrannical factorials, $\propto nN|\lambda N \propto nN - N$, for $N\pi 1$. This straightforwardly leads to the distribution:

$$N_v = Z_\nu \frac{1}{e^{\beta(E_v - \mu)} - 1}, \tag{5}$$

with

$$N = \sum N_v = \sum \frac{Z_\nu}{e^{\beta(E_v - \mu)} - 1}, \tag{6}$$

where $\beta = 1/k_B T$.

Note that the expression for Z_ν hides in it the single-particle density-of-states factor which depends on the dimensionality.

Several observations can now be made on Eqs. (5) and (6) that eventually gave the BEC [4]:

(a) The expression on the right-hand-side of Eq. (5) is singular for $E_v = \mu$. The singularity is, however, integrable for a 3D gas, but logarithmically divergent in 2D.

(b) The chemical potential must be negative, including zero, as N_v has to be a non-negative number.

(c) The chemical potential must increase towards zero as the temperature is lowered for a given N, while the right-hand-side of Eq. (6) decreases continuously.

(d) At a certain critical value T_c, the chemical potential vanishes and remains stuck at zero then on for lower temperatures.

(e) Below the critical temperature, the right-hand-side of Eq. (6) becomes less than N for a given T: the thermal distribution now cannot hold all of N bosons in the thermally excited non-zero energy states — there is an over-population!

(f) The excess population must necessarily accumulate in the singularity at the lowest single-particle state, i.e. the zero momentum state. This is how Einstein had argued, and was thus led to condensation, and to the condensate fraction. The above-the-condensate fraction remains as a saturated ideal gas (vapor) in equilibrium with the condensate.

The critical condition ($\mu = 0$) for the condensation is best expressed in terms of the phase–space density:

$$n\lambda_{\rm dB}^3 \geq 2.612\,, \qquad \lambda_{\rm dB} = h/(2Mk_BT)^{1/2}\,. \tag{7}$$

This is also called the condition for quantum degeneracy. The equality sign in Eq. (7) holds at the critical temperature. Here $\lambda_{\rm dB}$ is the thermal de Broglie wavelength, and $n = N/V$ is the number density. (This condition for BEC can be restated as that the mean inter-particle spacing be less than the de Broglie wavelength ensuring appreciable overlap of the thermal wavepackets, that makes the indistinguishability effective.)

It is clear from Eq. (6) that there is no BEC in a 2D gas where the density-of-states has a non-zero value at zero energy. This is unlike the case of a 3D gas where the density-of-states vanishes at the bottom, i.e. at the zero of the single-particle energy. It is the wholeness of the BEC (a single macroscopic object) that accommodates the excess plurality (population) in the single zero-momentum state. It is in this sense that there is plenty of room at the bottom.

Einstein clearly realized that BEC is a purely-quantum statistical effect. He did not refer to the condensation as a phase transition. Einstein, however, had the mental picture of the condensate fraction in equilibrium with the above-the-condensate fraction much as the saturated vapor is in equilibrium with the liquid phase under isothermal conditions. Though, in a BEC the phase separation is in the momentum space.

Several other things also followed naturally from his derivation: The Nernst Theorem was satisfied (entropy vanished at the zero of temperature as there was a single state — the BEC); the Gibbs paradox was obviated, without recourse to any *fixing* or *correction*, such as dropping the factorial $N!$ from the Boltzmann way of counting. This made the entropy correctly additive.

Note on indistinguishability

Two objects may be said to be indistinguishable if they are merely two different states of the same underlying entity. This, of course, happens naturally in a quantum-field description where the particles are the excitations of an underlying field — just its internal movements. At a somewhat heuristic level, one can understand the quantum indistinguishability of the classically identical particles. Classically, it is possible in principle to keep track of the identity of the particles as they are being permuted — here permutation is viewed as a process. The continuous tracking makes it always possible to know which is which. Quantum mechanically, however, there are no trajectories and thus keeping track of the identical particles is forbidden, in principle. Hence their indistinguishability. This is as operational as one can get. There remains, however, a question: is there a degree of indistinguishability, e.g. two particles differing arbitrarily slightly in their masses. Is approximate indistinguishability meaningful, or must it be an absolute condition? Some of these questions had occurred to Einstein [3] — in the form of a paradox involving a mixture of two gases with slightly differing molecular masses. Einstein also examined thermal fluctuations of the number in the Bose system (fluctuations, and, of course, invariances being his abiding interests). He found the wave noise (an interference effect) in addition to the shot noise-rediscovered many a time since.

Finally, a note on BEC in a spatially localized quantum gas. This is relevant to the BEC now realized in the optical and magnetic traps [6–9] Here the usual condition $Z_\nu \gg 1$ for the phase-space elements on an energy shell is clearly not satisfied. One must do the fully quantum treatment using the occupation number representation for the Bose system with its second quantized creation/annihilation operators [8].

Generalization of BEC

The BEC derived by Einstein was only for an ideal gas of non-interacting scalar bosons, extended uniformly in the three-dimensional (3D) space. Generalization has since been considered and in some cases realized:

(a) fragmented BEC [10–12] — extrinsically into non-overlapping regions of coordinate or momentum space; and intrinsically in terms of their internal (hyperfine) spin structures, or even their macroscopic quantum mechanical phases. Then, there is the question of their thermodynamic stability — repulsive interaction as in ^{4}He has been shown to disfavor fragmentation for reasons of the energetics of exchange interactions [11];
(b) Dimension — there is no BEC in two dimensions, as can be readily seen from the fact that the sum of the thermal occupation numbers (see Eq. (6)) over the molecular states for the system then diverges at all temperatures for chemical potential $\mu = 0$. The absence of BEC in two dimensions, of course, follows from a rather general theorem in condensed matter physics;
(c) Localized condensates [6–9] — BEC has now been realized in dilute alkali atomic gases in harmonic traps, magnetic and optical;

(d) Interactions [13] — interacting bosons, for example ^{4}He, has been treated extensively by the many-body theorists. Interactions (repulsive) deplete the condensate;
(e) Condensate fraction — neutron scattering [14] has been an experimental technique of choice for determining the condensate fraction in ^{4}He.

In neutron scattering with high energy-momentum transfer rates, the struck atoms of the condensed system are excited to energies much greater than the binding energy, and thus the scattered cross-section gives directly the momentum distribution (the so-called Compton profile) of the system. The macroscopic occupation of the zero momentum single-particle state (the BEC) should now show up as sharp delta-function singularity (peak) in the measured momentum distribution, which is, however, yet to be clearly seen. It is essentially a *deep inelastic scattering* in the context of condensed matter. A novel method to demonstrate that a BEC really exists is the technique of quantum evaporation [15]. Here, a collimated beam of phonons injected into the sample causes evaporation of the atoms from the sample in a single-excitation to single-atom process. The angular distribution of the evaporated atoms was then inverted to show that there indeed is an accumulation of the atoms in the zero-momentum state — a BEC.

Bose–Einstein correlation

Closely related to BEC is the phenomenon of Bose–Einstein correlation (bec), where bosons of the same kind emitted from nearby sources get correlated in energy (E) and momentum (P), or time (t) and space (x). This is best seen with reference to the scattering of the two Bose particles at an ideal 50 : 50 beam splitter as depicted in Fig. 2.

For the one-boson incoming state, we have

$$a_1^+|vac\rangle \to (ra_3^+ + ta_4^+)|vac\rangle\,,$$
$$a_2^+|vac\rangle \to (ra_3^+ - ta_4^+)|vac\rangle\,,$$

in obvious notation, where a_i^+ is the Bose creation operator for the ith channel, and r and $-t$ the (real) elements of the scattering matrix for the beam splitter.

Now, for the two-boson incoming state, we have

$$a_1^+a_2^+|vac\rangle \to (r^2(a_3^+)^2 - t^2(a_4^+)^2 - rta_3^+a_4^+ + rta_4^+a_3^+)|vac\rangle\,.$$

Thus, for our beam splitter with $r = t$, we have only the doubly occupied outgoing states inasmuch as a_3^+ and a_4^+ commute for bosons. (For fermions,

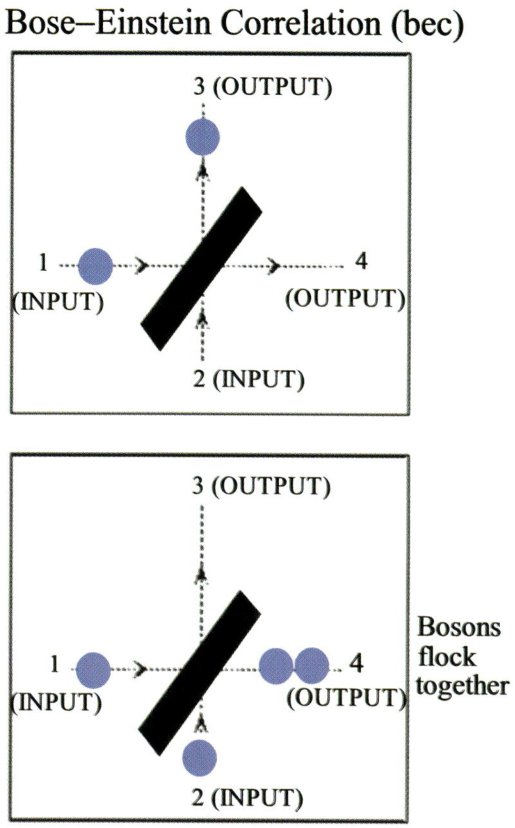

Fig. 2. Bosons emitted from nearby sources tend to be correlated in P and E or x and t.

of course, the reverse is true, and we will have only the singly occupied outgoing states.) This flocking of bosons of the same kind is, of course, crucial to BEC and to bosonic stimulation [16].

BEC and bosonic stimulation

This is closely connected with the $(1 + N)$ factor that multiplies the probability of a scattering event in which a Bose particle is scattered into a single-particle state that already has N bosons of the same kind. (This is, indeed, the 'crazy idea' that had intrigued Bose and Saha in the papers of Einstein and Ehrenfest [17] and of Pauli [18] written in the context of the probability of Compton scattering that depended on the radiation density at the scattering frequency that would arise in the process.) Bosonic stimulation is clearly involved in the kinetics of BEC growth, and may be used to amplify certain extremely weak effects in condensed matter.

Macroscopic wavefunction for BEC

Einstein had considered only the quantum statistics and the resulting thermodynamics of BEC. Fritz London [19] introduced the idea of BEC as a single macroscopic quantum object. The proximity of $T_{\text{BEC}} = 3.13$ K, calculated for ^{4}He regarded as an ideal Bose gas, and the observed $T_\lambda \varphi$ 2.18 K, led him to identify the lambda transition with the BEC. This eventually led to a nonlinear Schrödinger-like equation, the Gross–Pitaevskii equation [7, 8], for the macroscopic matter–wave function $\Psi(r)$, of a single coordinate r, describing the BEC in a realistic Bose system with interactions (e.g. ^{4}He):

$$i\hbar\frac{\partial}{\partial t}\psi_0(r,t) = -\frac{\hbar^2}{2m}\nabla^2\psi_0(r,t) + g|\psi_0(r,t)|^2\psi_0(r,t), g = \frac{4\pi\hbar^2 a}{m}, \tag{8}$$

with $\Psi_0(r,t)$ = Complex order parameter = $\sqrt{n_0(r,t)}e^{i\theta_0(r,t)}$, $n_0(r,t)$ = BEC number density; $\theta_0(r,t)$ = phase, v = velocity $\equiv (\hbar/m)\nabla\theta_0(r,t)$, where a is the s-wave scattering length that parametrizes the self-interaction ($a > 0$ for repulsive interaction).

Ψ_0 describes BEC

Space-time coherent phenomena — interference, diffraction
Superfluid flow through capillaries, past obstacles
Quantized vortices (h/m = quantum of circulation)
Tunneling through barrier
Nonlinear Matter Waves (4 wave mixing)
Collective excitations (the sounds).
At $T = 0$, the condensate fraction <1, but the superfluid fraction = 1.

BEC localized in traps

When Einstein derived the condensation for an ideal (non-interacting) gas of particles obeying the quantum (Bose) statistics, he had in mind helium, hydrogen (which are actually strongly interacting), and also the gas of electrons (wrongly, as electrons actually obey the Fermi–Dirac statistics, of course, not known then). A BEC with high condensate fraction requires a high phase–space density without having to encounter the adverse effects of strong interactions that not only deplete the condensate, but also actually pre-empt it by causing solidification. This suggests low densities and correspondingly low enough temperatures, and of course, low-mass atoms. This is precisely what has been achieved in the dilute gaseous alkali atom BECs, with typically ~10 orders of magnitude lower than the normal condensed matter density and a temperature ~100 nanokelvins. Thus, the domain of BEC has been extended far beyond helium, or shall we say, outside the helio-centric boundary in the laboratory. Extreme BECs are suspected in the cores of compact astrophysical objects, and in the cosmological vacua. Very recently [20], superfluidity (and by implication BEC) has been demonstrated in solid ^{4}He. Given below is the Zoo of BECs:

^{4}He (the inert noble, but Nobel-active gas)
H$\downarrow$ (spin-polarized hydrogen), an example of effectively spin-half Bose gas
Excitonic condensates
Composite bosons, e.g. $(e^- - e^-)$; (^{3}He–^{3}He) ...
Alkali atomic (bosonic) isotopes ^{87}Rb, ^{85}Rb, ^{7}Li, ^{23}Na, ...
Alkali molecules (fermionic-isotope pairs) $^{40}K_2$, $^6Li_2$
Protonic/neutronic and pion condensates — neutron star interior.
Cosmological condensates — field vacua.

Also, listed below are some parameter values typical of BECs in the laboratory for neutral bosonic alkali atoms:

Temperature	: 500 nk – 2 μk
Number density	: 10^{14} – 10^{15} cm^{-3}
Total number	: 10^3 – 10^7 – 10^9
Size and shape	: 10–50 μm spherical
	15 μm $\times$ 300 μm cigar shaped
Cooling cycle time	: few $\times$ seconds — few $\times$ minutes.

Open problems

Some of these are:

(a) Connection between BEC and superfluidity;
(b) Interaction and dimensionality;
(c) Fragmented BEC for composite bosons with internal structure;
(d) Kinetics of BEC growth;
(e) BEC and decoherence;
(f) Amplification of weak effects, e.g. the extremely small rotational magnetic moments expected of hydrogen molecules may add up coherently

to give a large macroscopic magnetic polarization in H_2 BEC!;

(g) Bosonic stimulation — one may even speculate about the decay rate of a radioactive nucleus being enhanced many-fold if embedded in the BEC of one of its bosonic decay products;

(h) BEC being a superfluid solid — a supersolid.

Timeline (fuzzy and annotated) of BEC

1900
- Planck's Quantum Hypothesis; Planck's Law of Black Body Radiation ('...happy guesswork').

1924
- Planck's Law and Light Quantum Hypothesis; Satyendra Nath Bose, *Zeit. f. Phys.*, 1924, **26**, 178. The pre-factor $2 \times (4\pi\nu^2/c^3)$ also derived.
- Beginning of Quantum Statistics: Loss of identity of light quanta. A new way of counting the indistinguishables.
- Photons indistinguishable but the phase space cells distinct.
- Photon number not conserved: Chemical potential $\mu = 0$

Quantum theory of the monatomic ideal gas, A. Einstein, *Preuss. Akad. Wiss.*, 1925, p. 3.

1925
- Extension of the Bose Statistics: Particle Number Conserved — Chemical potential $\mu \neq 0$, e.g. Helium (^{4}He).
- Gibbs paradox (the tyranny of N!) resolved, and the Nernst Theorem obeyed.
- Startling consequences: Macroscopic occupation of the lowest single-particle state — Bose–Einstein Condensation (BEC).
- Purely quantum statistical phase transition *sans* interaction.
- Initial reaction to BEC:
- Einstein ... '...that is only by the way ...', Planck ...frankly disbelieved it,
- Schrödinger ...suspected an error in it.

1926
- P. A. M. Dirac[21] gave antisymmetric wavefunction for fermions (^{3}He) obeying Pauli's exclusion principle with the occupation numbers restricted to 0 and 1 (the exclusive Fermi–Dirac statistics); and symmetric wavefunction for the bosons (^{4}He) obeying inclusive statistics with occupation numbers not restricted (the inclusive Bose–Einstein statistics). Matters of statistics were clarified by 1927.

1928 • W. Hendrik Keesom: He I (normal helium) – He II (superfluid helium) phase transition — the λ-transition at a critical temperature $T_c = 2.18$ K.

1938 • Pyotr Kapitza: Discovers superfluidity of helium (^{4}He). (Earlier in 1911 Heike Kamerlingh Onnes had discovered superconductivity in Sn).

• Fritz London Hypothesis: Superfluidity of ^{4}He a manifestation of BEC. A macroscopic wavefunction proposed for this phase. Now called the order parameter.

1940 • W. Pauli [22] derived spin-statistics connection from special relativity and quantum mechanics: bosons for integer spin and fermions for half-integer spins.

1948 • N. N. Bogoliubov [13]: First microscopic theory of interacting Bose-gas: ^{4}He — Superfluidity and BEC connected. Depletion of BEC due to strong interactions in liquid ^{4}He.

1956 • O. Penrose and Lars Onsager [23]: First estimation of BEC fraction ~10% for ^{4}He.

1957 • Bardeen–Cooper–Schrieffer (BCS) theory of superconductivity: Condensation of Bose-like Cooper pairs in the zero momentum state.

1966 • Seminal suggestion of P. Hohenberg and P. Platzman initiates probing of the condensate fraction by high-energy (epithermal) neutron scattering — momentum distribution (Compton Profile). But conflicting results for BEC fraction [14].

1972 *millikelvin* • Condensation of bosonic pairs of fermionic ^{3}He.

1980s *microkelvin* • Advances in laser cooling and trapping of neutral alkali atoms down to microkelvins; Steven Chu and William D. Phillips; and Claude Cohen–Tannoudji.

BEC SAGA: 70 years after 1925 and end of heliocentricity [6–8]

1995 • BEC RESURRECTED MIRACULOUSLY

∼nanokelvin Eric A. Cornell (NIST) Wolfgang Ketterle (MIT) and Carl E. Wieman (JILA +Univ. Colorado) obtain BEC in dilute gases of ^{87}Rb alkali atoms at ∼20 nK (0.00000002 K) and ^{23}Na. BEC fraction ∼100%, the ideal value.

∼1999 —
- New State of Matter: TUNABLE CONDENSATE
- Coherent matter waves–atom laser
- Bosonic stimulation
- Nonlinear matter — wave interaction: four-wave mixing (4 WM)
- Quantum phase transition: BEC in optical lattice
- Interaction tunable through Feshbach resonance

∼2003 —
- Fermionic atom pairs (Composite bosons): $^{40}K_2$, ^{6}Li molecular condensates.
- Close encounters: Cold collisions for scattering length $\ll$ de Broglie wavelength.

∼2004 —
- BEC (real-space pairs) — to — BCS (momentum-space pairs) crossover in fermionic systems
- Molecular BEC: Chemistry with cold coherent matter; Photo-association of atoms into molecules.
- Highest spatial and spectral resolutions; sensitive detectors (possibly for gravitational waves?).
- BEC on a microchip.
- BEC: A 'laboratory' for testing condensed matter models of strongly interacting systems, e.g. Mott insulator to superfluid transition.

References

[1] Bose, S. N. [1924] Planck's Law and light quantum hypothesis, *Z. Phys.* **26**, 178–181.
[2] Einstein, A. [1924] Ueber den Aether, *Verh. Schw. Naturf. Ges.* **105**, 85–93.
[3] Einstein, A. [1924] Quantentheorie des Einatomigen idealen gases I, *Preuss. Akad. Wiss.*, 261–267.
[4] Einstein, A. [1925] Quantentheorie des Einatomigen idealen gases II, *Preuss. Akad. Wiss.*, 3–14.
[5] For an insightful review of the ideas and the events bearing on Bose–Einstein Condensation, see Pais, A. [1982] *Subtle is the Lord* (Oxford University Press).

[6] Anglin, J. R. and Ketterle, W. [2002] Bose–Einstein condensation of atomic gases, *Nature* **416**, 211–217.
[7] Leggett, A. J. [2001] Bose–Einstein condensation in the alkali-gases: Some fundamental concepts, *Rev. Mod. Phys.* **73**, 307–356.
[8] Dalfovo, F., Stefano, G., Pitaevskii, L. P. and Stringari, S. [1999] Theory of Bose–Einstein condensation in trapped gases, *Rev. Mod. Phys.* **71**, 463–510.
[9] See Nature Insightful articles [2002] Ultracold matter, *Nature* **416**, 206–238.
[10] T.-L. Ho and S. K. Yip [2000] Fragmented and single condensate ground states of spin-1 Bose gas, *Phys. Rev. Lett.* **84**, 4031–4034.
[11] Noziéres, P. [1995] Some comments on Bose–Einstein condensation, in *Bose–Einstein Condensation*, eds. Griffin, A., Snoke, D. W. and Stringari, S. (Cambridge University Press).
[12] Van den Berg, M. and Lewis, J. T. [1981] On the free boson gas in a weak external potential, *Commun. Math. Phys.* **81**, 475–494.
[13] Bogoliubov, N. [1947] On the theory of superfludity, *J. Phys.* (*USSR*) **11**, 23–32.
[14] See, Silver, R. N. and Sokol, P. E. [1990] Superfluid helium and neutron scattering, in *Condensate Saga*, Los Alamos Science Summer.
[15] Wyatt, A. F. G. [1998] Evidence for a Bose–Eisntein condesate in liquid ^{4}He from quantum evaporation, *Nature* **391**, 56–59.
[16] Miesner, H.-J., Stamper-Kurn, D. M., Andrews, M. R., Durfee, D. S., Inaouye, S. and Ketterle, W. [1998] Bosonic simulation in the formation of a Bose–Einstein condensate, *Science* **279**, 1005–1007.
[17] Einstein, A. and Ehrenfest, P. [1923] Zur quantentheorie des strahlungsgleichgewichts, *Z. Phys.* **19**, 301–306.
[18] Pauli, W. [1923] Uber das thermische gleichgewicht zwischen strahlung und freien elektron, *Z. Phys.* **18**, 272–286.
[19] London, F. [1954] *Superfluids* (Dover, New York), vol. II.
[20] Kim, E. and Chan, M. H. W. [2004] Probable observation of a supersolid helium phase, *Nature* **427**, 225–227; Leggett, A. J. [1970] Can a solid be superfluid? *Phys. Rev. Lett.* **25**, 1543–1546.
[21] Dirac, P. A. M. [1926] On the theory of quantum mechanics, *Proc. R. Soc.* **112**, 661–677; also [1977] in *History of Twentieth Century Physics*, Varenna Summer School (Academic Press, New York).
[22] Pauli, W. [1940] The connection between spin and statistics, *Phys. Rev.* **58**, 716–722.
[23] Penrose, O. and Onsager, L. [1956] Bose–Einstein condensation and liquid helium, *Phys. Rev.* **104**, 576–584.

CHAPTER 8

Many Electrons Strongly Avoiding Each Other: Strange Goings On

****

T. V. RAMAKRISHNAN

Department of Physics, Banaras Hindu University, Varanasi-221 005, India

Indian Institute of Science, Bangalore-560 012, India

Our ideas about the behavior of a collection of large number of electrons in solids are based on regarding them as a quantum gas or fluid. This enables us to make sense of much of solid state science. However, an increasingly large number of families of systems where electrons move not freely but are subject, for example, to very strong short range (effective) repulsion, continue to be discovered and explored. Examples are high temperature superconducting cuprates, colossal magnetoresistance manganites and rare earth heavy fermion intermetallics. These systems exhibit a rich variety of qualitatively new properties which, perhaps, require fundamentally different ideas for many electron behavior of strongly correlated systems. I describe here some examples of such strange goings on; this is one of the major contemporary departures in the physics of condensed matter.

1. Introduction

A few years before 1905, the miraculous year whose centenary is being marked as the World Year of Physics, P Drude proposed (in 1900) a staggeringly simple model for electrons in solids [1]. Just three years earlier, it had been recognized (J. J. Thomson) that electrons are constituents of all matter. Drude suggested that the outer electrons of the atoms/molecules constituting the solid be regarded as forming an ideal, perfect, gas of charged particles. The broad reason is that each of these outer electrons is strongly influenced by the other atoms nearby, and so cuts loose from its parental ionic mooring (becomes unbound) and can move throughout the entire solid. With this simple model, Drude was able to describe well, for the first time, many observed (electromagnetic) properties of metals (e.g. their optical conductivity and 'skin' effect).

Two important reasons why this classical ideal electron gas description is basically inadequate, are the following. Firstly, electrons are actually waves, and their typical wavelength (thermal de Broglie wavelength) is comparable to the characterristic distance between them in solids, so that the electron gas is highly quantum mechanical; it needs to be thought of as a collection of a large number of independent *waves* satisfying the Pauli exclusion principle (namely there is only one electron in one state, the latter being characterized by the propagation vector of the electron wave and its spin direction). The quantum energy scale ϵ_F for typical electron densities in solids corresponds to a temperature of a few ten thousands of degrees, namely $\epsilon_F = k_B T_F$ with $T_F \sim 5 \times 10^4$ K, say. Thus at temperatures normally accessible to us, e.g. $0\text{ K} \lesssim T \lesssim 10^3$ K, one is deep in the quantum regime, $T \ll T_F$. This leads to novel universal electronic properties common to all such solids (metals), e.g. specific heat linear in temperature and paramagnetic (spin) susceptibility independent of temperature. This is indeed observed and contrasts with expectations from the classical Drude model, of temperature independent specific heat (Dulong Petit law) and T^{-1} (Curie) behavior for spin susceptibility. The electromagnetic properties as calculated by Drude do not change.

Secondly, the electron waves move not in a homogeneous, neutralizing, dynamically inert background, but in the background of ions which (for a crystal) are arranged in a spatially periodic lattice. This leads to states with energies of progressive electron waves being organized in bands separated by forbidden gaps. There is then the possibility of semiconductors or insulators in addition to metals, the former being realized when the number of unfilled shell electrons is such that the (highest allowed) energy band is completely full, and is separated from the lowest energy unfilled band state by an energy gap.

Much of our knowledge of electronic behavior of solids is based essentially on these ideas, and on treating other physical effects as perturbations. Such effects arise for example from the fact that the ions are not fixed but vibrate (phonons), or from disorder in their spatial arrangement; electrons also repel each other via electrostatic or Coulomb interactions so that they cannot be regarded as moving independently of each other. Each of the above effects if strong, is known to lead to qualitatively new electronic behavior. Here we focus largely on the last.

Coulomb interactions between electrons are clearly always present and can be strong. A natural and simple approximation is to consider each electron as moving in a self-consistently determined average static potential or mean field due to the others. A sophisticated version of this, due to Landau (1954) [2], describes the elementary low lying excitations in metals as well-defined quasiparticles interacting with each other. The quantum numbers of these quasiparticles are in one to one correspondence with those of free fermions, e.g. having wavevector $\vec{k}$ and spin σ. Interaction determines the excitation energy $E_{\vec{k}\sigma}$ of a quasiparticle, and the coupling $f_{\vec{k}\sigma,\vec{k}'\sigma'}$ between them. This is clearly a picture of the low energy excitations of the interacting Fermi system (commonly called a Fermi liquid) as a true image of the noninteracting one. The fundamental reason as to why Landau's Fermi liquid theory might be realistic is that the Pauli exclusion principle strongly constrains the transitions from a given state that can be caused by interactions; for small excitation energy ϵ with respect to the chemical potential μ (namely the surface in wavevector space with zero excitation energy i.e. the Fermi surface) the decay rate $\Gamma(\epsilon) \propto \epsilon^2$, so that crudely $\Gamma(\epsilon) \sim (V_{\text{int}}^2/\mu^3)\epsilon^2$ where V_{int} is the interaction strength. Clearly, for small enough $\epsilon, \Gamma(\epsilon) \ll \epsilon$ no matter how large is V_{int}, so that low excitation energy states are long lived and well defined. The ideas involved in the Landau Fermi liquid theory are thus that there is a continuity in the states or state quantum numbers between the independent and the interacting fermion systems and that the effect of the interaction (e.g. on Γ) can be at least formally thought of perturbatively.

In 1937, de Boer and Verwey found and pointed out that a number of transition metal oxides which might to be metallic in band theory (i.e. because the number of electrons in each unit cell of the appropriate crystalline solid is an odd integer) are actually insulators. Peierls and Mott realized immediately that this fundamental difference could be because of strong interaction between electrons which forces them to 'stay home' with their parent atoms. Thus if the number of electrons per unit cell is integral (even or odd), the system is an insulator, since each electron is localized by correlation. In 1949, Mott proposed a simple hypothetical model, of a lattice of hydrogen atoms. As its lattice constant increases, the system goes from being a half-filled band metal to an insulating collection of hydrogen atoms. This metal to insulator transition is called a Mott transition, and the electron correlation driven insulator, a Mott insulator. Strong correlation causes here a qualitatively new phenomenon, the localization of electrons. Experimentally, several families of solids with strong electronic correlations are known; many are being discovered and explored. All are strange, being quite different in their physical properties from 'Fermi liquid' type of systems we have been familiar with. I briefly describe three families below, while outlining the challenges they pose.

I do not mention several families of strongly correlated electron systems e.g. sodium cobaltate Na_xCoO_2 (of great current interest as home to a Curie–Weiss metal, for unusual charge ordering and superconductivity as well as possible realization of naturally frustrated triangular lattice systems with large and the quantum effects), spin chains and spin ladders. The great success story of fractional quantum Hall effect (FQHE) has not been touched on. The FQHE is the defining characteristic of an incompressible quantum fluid which owes its *existence* to (strong) interactions between electrons residing in the lowest Landau level. This novel quantum fluid in addition to exhibiting quantization of Hall conductance, has fractionally charged excitations (e.g. charge $e/3$) which have been observed. This is one example of a many electron system determined by interactions and understood in several ways, e.g. ground state wave function and excitations as well as effective low energy quantum field theory.

2. Cuprates

Perhaps the best known family of oxides in which strong correlation between electrons defines their low energy behavior is rare earth cuprates (3). The present period of intense interest and activity dates to 1986, when high temperature superconductivity was discovered by Bednorz and Muller in $La_{2-x}Sr_xCuO_4$ with $x \sim 0.2$. Nearly two decades and almost a hundred thousand papers later, we have a richly detailed experimental picture of their properties. There is, however, no agreement on the underlying theoretical description. There is direct experimental evidence e.g. from core photoemission that correlation energy U (effective onsite electron repulsion) is rather large, of order 5 eV or so, in these systems whereas the kinetic energy is much smaller, the intersite hopping energy t_{ij} between nearest neighbors i and j is about 0.3 eV. (The nominal bandwidth is $D = 2zt$ where z the number of nearest neighbors is 4 for a square lattice; thus in these systems, U is much larger than D, a signature of strong correlation.)

The parent compound La_2CuO_4 was studied by Ganguly and Rao in 1980 or so. It is a nearly two-dimensional solid, with the electronically

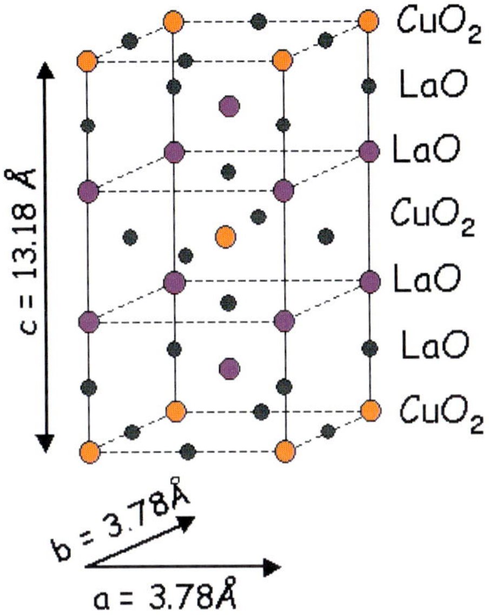

Fig. 1. Crystal structure of La_2CuO_4. A unit cell is shown. The ions are not to scale. (http://hoffman.physics.harvard.edu/research/SCmaterials.php)

active $Cu^{++}(d^9)$ ions forming a nearly square planar lattice with each Cu^{++} surrounded by four O^- ions; the (CuO_2) planes are interspersed with $(La\text{-}O)^+$ planes (Fig. 1). The *sp* electrons of Cu, O and La are strongly bonded; the low energy excitations of the solid involve the $d_{x^2-y^2}$ orbitals of the d state in Cu, appropriately hybridized with the surrounding oxygen atom states. These electrons have relatively weak interplanar coupling. Ganguly and Rao found that La_2CuO_4 is an antiferromagnetically ordered insulator; the antiferromagnetic coupling between nearest neighbor Cu^{++} spins in plane is estimated to be about 0.15 eV, an unusually large value. Above the Neel temperature of about 210 K, the paramagnet continues to be insulating. This is the signature of a Mott insulator, since there is an odd number of electrons in each unit cell containing one Cu^{++} ion, namely, one d hole or one $d_{x^2} - y^2$ orbital electron for each Cu^{++}.

Replacing trivalent La with a divalent alkaline earth effectively removes a $d_{x^2-y^2}$ electron, so that one has a doped Mott insulator, which is metallic because of hole motion. Bednorz and Muller found in 1986 that such a compound is a superconductor with an unprecedentedly high transition temperature; in the $La_{2-x}Sr_xCuO_4$ family the highest T_c is about 30 K, and

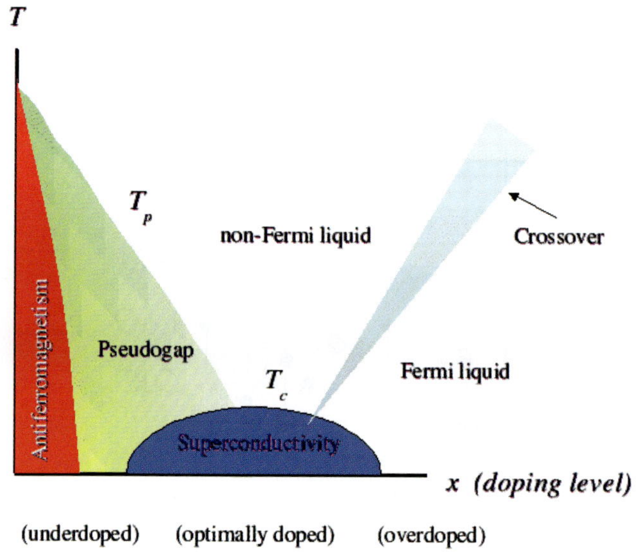

Fig. 2. The generic, schematic phase diagram of hole doped cuprates as a function of hole concentration x and temperature T. (Courtesy: C. M. Varma)

occurs for $x \sim 0.20$ ('optimum doping'). A number of cuprate families are known; the highest T_c observed is about 140 K. The generic 'phase diagram' of cuprates as a function of doping x and temperature T is shown in Fig. 2. Two kinds of ordered phases, namely antiferromagnetic insulator for very low doping, and a dome shaped superconducting region are shown. The metallic or 'normal' regime outside these can be of several types, there is a smooth crossover between them. The three large regions are characterized as pseudogap, non-Fermi liquid and Fermi liquid.

A qualitatively new feature of the metallic state (both the pseudogap and the non-Fermi liquid) is that an electron of a definite momentum $\vec{k}_{\|}$ (i.e. $\vec{k}$ in the plane) does not have a definite energy, but has a broad spread of energies. There are *no* well-defined quasiparticles, in stark contrast to expectations from Landau's Fermi liquid picture where it is believed that no matter how strong the interaction, electronic excitations with $\vec{k}$ close to the Fermi surface ought to be well defined. Recent advances in angle resolved photoemission spectroscopy (ARPES) enable very accurate measurement of the energy distribution of electrons with a well-defined momentum ejected from the solid by an incident photon. The expected ARPES 'spectrum' is

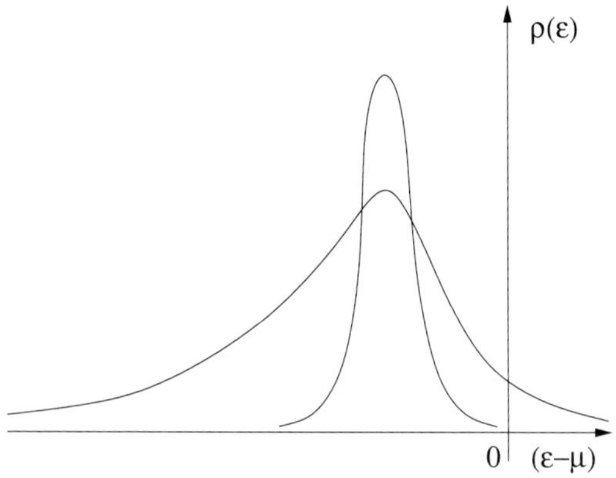

Fig. 3. Schematic ARPES intensity spectrum $\rho(\epsilon)$ as a function of energy $(\epsilon - \mu)$, for a particular planar electron wavevector.

a rather narrow peak. This is *not* what is seen (Fig. 3). The absence of well-defined quasiparticles marks cuprates out as a qualitatively new kind of many electron system. Intriguingly, a quasiparticle peak returns in the superconducting state; its strength is proportional to the superfluid density.

The cuprates are characterized by a host of unusual properties (3). We mention a few here. It has been noticed for long that in-plane resistivity in the non-superconducting, 'normal', phase is basically linear in T over an unusually wide range, roughly from T_c to 10^3 K or so. In conventional metals, there is a quantum energy scale ϵ_o for the resistive (bosonic) scatterers which sets the temperature above which $\rho(T) \propto T$, with a clear tendency for $\rho(T)$ to saturate as temperature increases to high values. In cuprates, this language would imply that ϵ_o is very small, nearly zero and that there is no tendency for resistivity saturation. Again unlike other metals, the electronic specific heat is not linear in temperature. The underdoped $(x < x_{\text{optimum}})$ region is electronically very unusual. There is a pseudogap regime above T_c. Here, the single particle excitations as seen from ARPES show a soft gap about the Fermi energy; the gap has fourfield symmetry in wavevector or $\vec{k}$ space, similar to that of the superconducting gap $\Delta_{\vec{k}}$. The latter fourfold symmetry feature specifically the k dependence $\Delta_{\vec{k}} \simeq \Delta_o(\cos k_x a - \cos k_y a)$ for $\vec{k}$ in the $x - y$ plane, is characteristic of cuprates; they are 'd-wave' superconductors. We notice that the gap *vanishes* at the points $k_x = \pm k_y$ on the Fermi surface. The strongly underdoped region is seen to have coexisting real space order $(4 \times 4$ superstructure) and superconductivity. The pseudogap reduces the density of low energy charge and spin excitations; this is seen, for example, in the characteristic decrease of low temperature resistivity below the linear dependence, and in the spin susceptibility (spin gap).

The attempts to make sense of electronic properties of cuprates range from assuming that they are ordinary metals but with a new kind of 'glue' that helps form electron pairs, to models with radically new elementary excitations. None has the comprehensive explanatory power to successfully confront the large number of novel experimental features. I briefly mention here the strong correlation viewpoint. It is based on the realization, as mentioned above, that the undoped cuprate is a Mott insulator with effectively one d electron (with spin 1/2) locked to each site, and with antiferromagnetic (AF) coupling J between such spins at nearest neighbor sites. On hole doping, an electron can find itself next to a site with a hole, and can hop on to it so that the system can become metallic because of the mobility of electrons (or of holes). Nearest neighbor electrons form singlets due to the coupling J. The superconductor is a coherent superposition of singlet pairs. Making a reliable theory out of such ideas proves difficult, essentially because it is hard to weave strong *local* constraints (namely that low energy states have exactly one or no electron per site) into global many fermion quantum dynamics. Thus the origin of the different non-Fermi liquid, nonsuperconducting metallic regimes, i.e. the question of how exactly do strong correlation, exchange J_{ij} and electron hopping t_{ij} lead to the observed states is not clearly answered; indeed there are several theories which argue for other interactions and degrees of freedom as being necessary.

3. Manganites

Alkaline earth doped rare earth manganites, namely $Re_{1-x}Ak_xMnO_3$ where Re is a rare earth ion e.g. La, Pr, Nd and Ak is an alkaline earth ion e.g. Sr, Ca, Ba ... are a family of structurally simple (perovskite, basically cubic) oxides with a rich variety of unusual phenomena and phases [4]. The best known of their properties is colossal magnetoresistance, namely the large change in electrical resistivity in a magnetic field. This change, several orders of magnitude larger than in typical metals, is one of a large number of properties which can be described as being due to the persistent proximity of the metallic and the insulating states. There is no general agreement on the origin of their unusual behavior. I mention below some characteristic

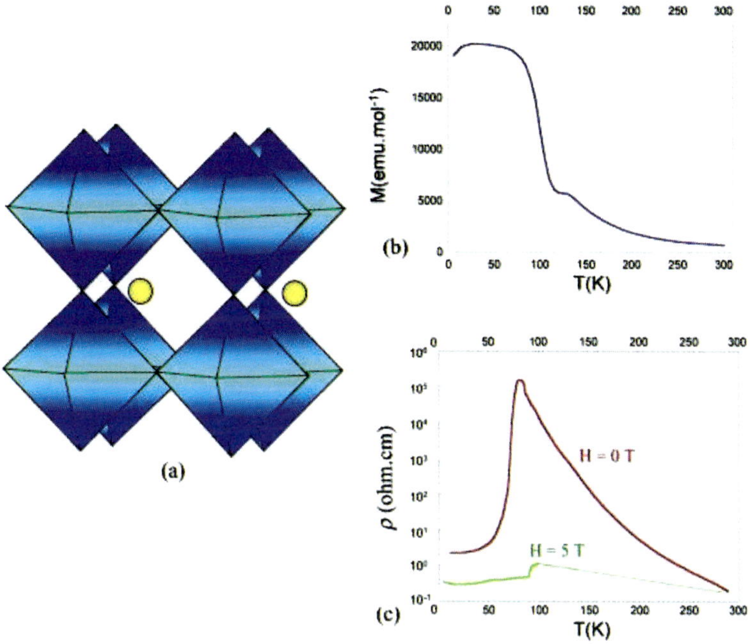

Fig. 4. (a) Corner sharing Mn-O_6 octahedra with 'interstitial' ($Re_{1-x}A_x$) ions. (b) Spontaneous ferromagnetic magnetization M as a function of temperature T, and (c) dependence of resistivity with temperature in zero magnetic field and in a field of 5T, both for a ceramic manganite.

properties, after a brief description of their solid state electronic structure. The kind of strong correlation effects which could be relevant, and a two electron fluid model proposed by us are then mentioned.

The manganites are made up of corner sharing octahedra. The Mn ion is at the center of octahedron and the oxygen ions are at the corners. The rare earth/alkaline earth ions are located between the octahedra, in the interstitial spaces [Fig. 4(a)]. Because of the octahedral coordination, the d electron states, which are five-fold orbitally degenerate in the free atom (isotropy) split into three-fold degenerate t_{2g} levels and two-fold degenerate (e_g orbitals the two levels are denoted $d_{x^2-y^2}$ and $d_{3z^2-r^2}$). The compound $Re_{1-x}Ak_xMnO_3$ has a fraction $(1-x)$ sites with Mn^{3+} ions $(t_g^3 e_g^1)$ and x sites with Mn^{4+} ions $(t_{2g}^3 e_g^o)$. The t_{2g} levels are thus always occupied at every Mn site by three electrons and can be thought of spins as $\vec{S}_i$. The two-fold degeneracy of the e_g level is spontaneously broken by a

Jahn–Teller distortion of the (Mn-O_6) octahedron on sites where one electron occupies the e_g state. In manganites, the Jahn–Teller energy lowering E_{JT} is large ($\sim$ 1 eV), as is the corresponding Jahn–Teller polaronic distortion, since the relevant electron lattice coupling λ is strong. The strong local Coulomb repulsion between electrons in the Mn d orbital has at least two effects. One is described phenomenologically by Hund's rule; in the ground state the t_{2g} spins have their maximum value i.e. $\vec{S}_i = 3/2$ and the e_g and t_{2g} spins at a site tend to be parallel. This last is described by an effective ferromagnetic coupling $-J_H \vec{s}_i \cdot \vec{S}_i$ with $J_H \simeq 2$ eV where $\vec{s}_i$ is the e_g spin at site i. The second effect is the effective repulsion U between e_g electrons. U is estimated to be about 5 eV. Thus in say $La_{0.8}Ca_{0.2}MnO_3$, e_g electrons (in a mixture of two orbital states) hop from site to site under the influence of *three strong* local interactions, namely coupling to lattice (strength λ), Hund's rule coupling to t_{2g} spins (J_H) and electron repulsion U. The intersite hopping energy has a scale $t \sim 0.2$ eV with a factor of order unity which depends on the initial and final orbitals. This system is thus strongly correlated since $t \ll E_{JT}$, J_H and U. Orbital, lattice and spin degrees of freedom are explicit and strongly coupled. If the local interaction were to have no qualitative effect, the result of hole doping would be to enable the e_g electrons to move through the lattice; electronically the hole doped manganite would be a metal. In fact, this is not true over a wide range of doping x so that the local interactions have qualitatively new effects.

The manganites exhibit a variety of orbital, magnetic and structural order; electronically both metallic and insulating states are found. As expected from the above description, there is an intimate connection between them all. An example is colossal magnetoresistance (*cmr*), which is found near the Curie or paramagnetic–ferromagnetic transition, which also nearly coincides with an insulator-metal transition [Fig. 4(b)]. Figure 4(c) shows the large reduction in resistivity near the Curie transition due to a 5T magnetic field. Figure 5 shows the magnetic and structural phases in $La_{-x}Ca_xMnO_3$ for different doping x. One has different kinds of AF order in the insulating ground state, as well as ferromagnetic insulating (FI) and ferromagnetic metallic (FM) ground states. There is no orbital order for $0.2 \lesssim x \lesssim 0.5$, (orbital liquid regime) but orbital order with related structural long range order exists for $x \gtrsim 0.5$. While the complexity of the system and the presence of many interactions could be the reason for the richness of phase diagram, other manganites show similar phases with systematic differences.

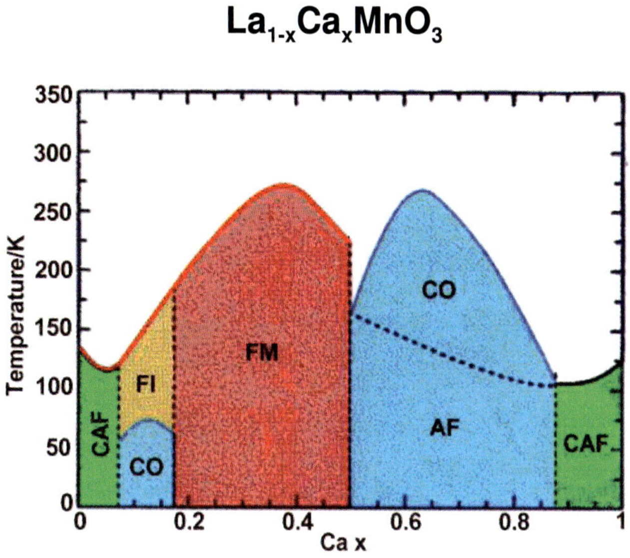

Fig. 5. Phase diagram of $La_{1-x}Ca_xMnO_3$; different structural and magnetic phases are shown. The symbols stand for the following: R (rhombohedral) O*, O (orthorhombic) CAF (C type antiferromagnet), CO (charge order), FI (ferromagnetic insulator), FM (ferromagnetic metal), AF (antiferromagnet, insulating). (S. W. Cheong and H. Y. Hwang, in *Collossal Magnetoresistive Oxides*, ed. Y. Tokura (Gordon and Breach, Amsterdam, 2000))

As mentioned earlier, a special characteristic of manganites is the fact that metallic and insulating electronic configurations seem close in energy over a wide range of composition and temperature. We mention some illustrative examples. Such proximity is peculiar since electronic states near the Fermi energy need to be either localized, or be extended and gapped, for an insulator. The opposite is true for a metal. So, generally a physical system is one or the other, and only for special conditions (e.g. of temperature, pressure or composition) does a transition takes place. By contrast, e.g. in $La_{-x}Ca_xMnO_3$, there is an insulator to metal transition on cooling, over the *entire* composition range $0.2 \lesssim x \lesssim 0.5$. Colossal magnetoresistance itself implies that a magnetic field is enormously effective in 'metallizing' the system. A manganite compound $(La_{1-y}Pr_y)_{0.7}Ca_{0.3}MnO_3$ at $y = 0.75$ is poised so close to the transition that replacing O^{18} by O^{16} changes it from insulator to metal. Another interesting feature of manganites is that the insulating, charge and orbitally ordered state ($x \gtrsim 0.5$) 'melts' in a relatively small magnetic field into a metal. Finally, there is a large amount of direct experimental evidence from various local probes that two electronically very different regions, namely metallic with no local lattice distortion, and insulating with local lattice distortion, coexist on length scales ranging from nanometers to microns; the regions can be static or dynamic. One of the major questions in this field is whether these electronic inhomogeneities are extrinsic or intrinsic, the origin of their length and time scales, and whether this is a defining characteristic ('electronic softness') of strongly correlated systems.

While from the description above it is apparent one might attribute the unusual and varied properties of manganites to strong local interactions, theoretical descriptions take these into account selectively, perhaps because only some (depending on the experimental conditions) could be qualitatively significant. For example, large Hund's rule coupling in tandem with e_g electron hopping leads to a new effective ferromagnetic coupling (double exchange) between nearest neighbor e_g spins whose strength is roughly proportional to $(1-x)t$, $((1-x)$ being the fractional number mobile electrons and t the electron hopping) times a spin $1/2$ overlap factor which decreases further from unity as the nearest neighbor t_{2g} spins become less parallel. The effect of the large Jahn–Teller electron phonon coupling is to create a polaron (electron with associated local lattice distortion) states. In the static or classical approximation for the lattice displacements these states continue to form a broad band. In the presence of large U, the effective bandwidth of the e_g electrons is proportional to the hole density x. It could be assumed that this effect renormalizes the bare e_g bandwidth and also has no qualitative consequences. Recently, some of us (T. V. Ramakrishnan, H. R. Krishnamurthy, S. R. Hassan and G. V. Pai) [5] have proposed a strong correlation two fluid model as a basic theoretical picture for manganites. We argue that because of strong Jahn–Teller coupling, there is a JT polaron (denoted ℓ_i) with site energy lowered by E_{JT} whenever there is an e_g electron on that site (as pointed by Millis, Littlewood, Mueller and Shraiman, who also were the first to analyze the significance of polarons for mangnite phenomena).

There is also at each site, an othogonal state b_i. The intersite hopping of the ℓ polaron is reduced by a factor $\sim \exp(-E_{JT}/2\hbar\omega_o)$ where $\hbar\omega_o$, the Jahn–Teller phonon energy $\hbar\omega_o \sim 0.05$ eV. Since this factor is $\sim (1/200)$, the ℓ polaron fluid is essentially site localized (this is a consequence of phonon dynamics; it is absent if lattice displacements are treated classically).

The b electron hops from site to site without any polaronic reduction in bandwidth, avoiding polaronic sites because of the large onsite repulsion U between ℓ and b states. Thus of the three strong onsite interactions, one (the electron lattice coupling) leads to two fermion species, an ℓ polaron which is essentially localized with site energy $-E_{JT}$ and a band electron (b). The strong correlation U and the Hund's rule coupling J_H affect ℓb dynamics. In a strong correlation, single self-consistent site (dynamical mean field theory, DMFT) approximation, we show that many of the observed novel phenomena in the orbital liquid regime are due to the relative occupation of the b band and the ℓ level. The system is *not* a Fermi liquid. Since the b bandwidth effectively decreases with decreasing x (an effect of strong correlation U) and decreases with increasing temperature T (effect of Hund's rule coupling J_H) so does the average b electron number $\bar{n}_b$. Using this idea, metal insulator transitions as a function of x and T, and interestingly colossal magnetoresistance, the smallness of carrier density (in our model the b electron density) can all be physically and quantitatively explained. A number of phenomena (in this strong correlation model) depend on spatial correlation between orbitals, ℓ polaron hopping, disorder, etc. Theoretical description of these is to be developed. The approach however shows that strong coulomb correlation and electron lattice effects in these systems crucially determine the physics and act in concert.

4. Heavy Fermions

Intermetallic compounds containing rare earth ions, e.g. $CeA\ell_3$, can be described electronically as of a collection of f and *spd* electrons. The former continue to be 'attached' to their parent atoms while the latter are free electrons spread throughout the solid. The f electron number in each atom is integral (strong correlation limit) and thus each atom has a magnetic moment. The f electron also hybridizes weakly with the conduction or *spd* electrons. The presence of strongly correlated localized states which mix with band states leads again to several new kinds of phenomena in those systems. Anderson showed, in a famous paper on localized moments in metals (1957) that if the local correlation energy U is, roughly, larger than the decay rate of the local state due to hybridization, the local state has a moment in a (mean field) theory in which the average effect of conduction electrons is considered. But it is clear that a moment in the sense of a local rotational symmetry breaking state cannot survive quantum fluctuations

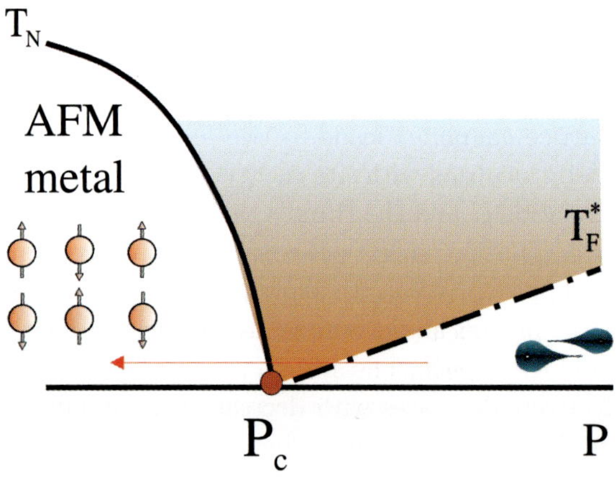

Fig. 6. Schematic plot of the temperature T, pressure P phase diagram of a typical heavy fermion system. P_c is the quantum critical point. For $P < P_c$, the ground state is shown to be an antiferromagnetically ordered metal. For $P > P_c$, it is a paramagnetic metal without magnetic moments. The (generally) non-Fermi liquid metal exists above the Neel temperature T_N for $P > P_c$ and the crossover 'heavy fermon' temperature T_F^* (shown by the dot dashed line) for $P > P_c$. (Courtesy: P. Coleman)

involving the residual coupling of the local moment with conduction electrons which restores notational symmetry; this is the Kondo effect. The Kondo temperature scale is exponentially small. In a lattice of f electron atoms, the moments interact, and the f electrons move through the lattice via the conduction electrons. They thus effectively form a narrow band of heavy fermions. Effective mass values a few hundred times the free electron mass have been observed. The heavy fermions (basically f electrons) form a Fermi system with a characteristic Fermi temperature which ranges from 0 to 10^2K! There is a competition between coupling among magnetic moments which tends to order them spatially (e.g. antiferromagnetically), the Kondo effect which tends to 'quench' each moment locally, and, heavy fermion behavior. One consequence of this competition is that as a function of some control parameter (e.g. pressure) the ground state may change from that with moments and long range (antiferro) magnetic order to one in which moments disappear and the system is a nonmagnetic metal, at a quantum critical point (QCP) (Fig. 6). The domain at nonzero T around the QCP, without long range order, is a novel kind of metal, often seen to be a non-Fermi liquid. It is a special kind of strongly correlated system which is home to a novel kind of competition between spatially local magnetic fluctuations which restore rotational symmetry and interactions between magnetic moments which are consequences of strong correlation. Experimentally, there is a wide variety of systems in this family. The phenomena can also be tuned by magnetic fields in addition to standard methods such as pressure and composition, and can be thought of as arising from the interplay of coupled bosonic excitations involving moments and low energy fermionic excitations. There is great ferment, both theoretical and experimental, in this field, with proposals for novel kinds of interacting quantum field behavior and topological effects, e.g. for QCP and related phenomena, as well as unexpected experimental discoveries [6–9].

5. Conclusion

I have mentioned above several families of many electron systems in which quite likely because of strong local interaction between electrons, novel kinds of phenomena occur. No comprehensive paradigm, akin to the free electron gas/Fermi liquid theory exists. Perhaps there is no single/simple way in which local ('gauge') constraints and global quantum many fermion dynamics can be integrated. A number of quantum field theoretical models have been proposed and investigated. These theories open our eyes to many new kinds of ground states, order, excitations and correlations possible and are often realized as some of the 'quantum complexities' of condensed matter. However, I have tried to focus on a few broad classes of strongly correlated systems whose rich phenomena are not yet comprehensively understood. The ingredients are often known, and there are very many attempts to explore theoretically these systems in terms of known ideas suitably adapted, or in terms of novel interacting field models. They do not often seem to meet the essential experimental realities; there is a need for appropriate level strong correlation many body theories. The situation was described by Anderson some years ago as a 'logjam in condensed matter physics'. While there is exciting progress in our ability to experimentally design and probe new quantum worlds on the atomic scale, a number of these worlds are well-described theoretically, and this interaction continues to be creative, I have tried to make the point that the core problem of strongly correlated electron systems continues to be experimentally rich with unexpected discoveries, perhaps requiring new approaches.

It is thus specially appropriate to recall this area as we celebrate the world year of physics.

References

[1] See a textbook such as Ashcroft, N. W. and Mermin, N. D. *Solid State Physics* (Holt, Rinehart and Winston, New York, 1976), or Kittel, C., *Introduction to Solid State Physics* (Wiley, New York, 2005).

[2] See e.g. Baym, G. and Pethick, C. J., *Landau Fermi-Liquid Theory: Concepts and Applications* (Wiley, New York, 1992).

[3] A series of review articles in *Reviews of Modern Physics* on various aspects of high temperature superconductors from different points of view has appeared over the last few years. Some are the following: (i) Lee, P. A., Nagaosa, N. and Wen, X.-G., Doping a Mott insulator: Physics of high-temperature superconductivity **78**, 1 (2006); (ii) Basov, D. N. and Timusk, T., Electrodynamics of high-T_c superconductors **77**, 721 (2005); (iii) Hussain, Z., Shen, Z.-X. and Damascelli, A., Angle-resolved photoemission studies of the cuprate superconductors **75**, 473 (2003).

[4] See for example *Colossal Magnetoresistive Manganites*, ed. Chatterji, T. (Springer, Heidelberg, 2004); Jaime, M. and Salamon, M. B., *Rev. Mod. Phys.* **73**, 583 (2001).

[5] Ramakrishnan, T. V., Krishnamurthy, H. R., Hassan, S. R. and Pai, G. V., *Phys. Rev. Lett.* **92**, 157203 (2004) and cond-mat 0308396.

[6] See for example *Magnetism of Heavy Fermion Systems*, ed. Radowsky, H. B. (World Scientific, Singapore, 2002).

[7] Schroder, A. *et al.*, Onset of antiferromagnetism in heavy fermion metals, *Nature* **407**, 351 (2000).

[8] Coleman, P., Condensed-matter physics; magnetic spins that last for ever, *Nature* **413**, 788 (2001).

[9] Coleman, P. and Schofield, A., Quantum criticality, *Nature* **433**, 226 (2005).

CHAPTER 9

Einstein and the Quantum

****

VIRENDRA SINGH

Tata Institute of Fundamental Research
Homi Bhabha Road, Mumbai 400 005, India

We review here the main contributions of Einstein to the quantum theory. To put them in perspective, we first give an account of Physics as it was before him. It is followed by a brief account of the problem of black body radiation which provided the context for Planck to introduce the idea of quantum. Einstein's revolutionary paper of 1905 on light-quantum hypothesis is then described as well as an application of this idea to the photoelectric effect. We next take up a discussion of Einstein's other contributions to old quantum theory. These include (i) his theory of specific heat of solids, which was the first application of quantum theory to matter, (ii) his discovery of wave-particle duality for light and (iii) Einstein's A and B coefficients relating to the probabilities of emission and absorption of light by atomic systems and his discovery of radiation stimulated emission of light which provides the basis for laser action. We then describe Einstein's contribution to quantum statistics viz Bose–Einstein Statistics and his prediction of Bose–Einstein condensation of a boson gase. Einstein played a pivotal role in the discovery of Quantum Mechanics and this is briefly mentioned. After 1925, Einstein contributed mainly to the foundations of Quantum Mechanics. We choose to discuss here (i) his Ensemble (or Statistical) Interpretation of Quantum Mechanics and (ii) the discovery of Einstein–Podolsky–Rosen (EPR) correlations and the EPR theorem on the conflict between Einstein-Locality and the completeness of the formalism of Quantum Mechanics. We end with some comments on later developments.

1. Physics before Einstein

Albert Einstein (1879–1955) is one of the two founders of quantum theory along with Max Planck. Planck introduced the 'quantum' of energy in his investigations of black body radiation in 1900. He was followed by the young

Einstein who proposed the 'light quantum hypothesis' in 1905. Albert Einstein sent his revolutionary "light quantum" paper for publication on 17 March 1905 to *Annalen der Physik*. He was twenty six years of age and it was his first paper on quantum theory. He had published five papers earlier during 1901–1904 in the same journal. Those dealt with capillarity and statistical mechanics. The major frontier areas of research in physics then were thermodynamics and electrodynamics. The main conceptions about the physical universe prevalent in physics of that time were as follows.

1.1. *'Newton's mechanical conception'*

The earliest of these was that of a "mechanical universe" given by Isaac Newton in his magnum opus "Principia" in 1687. The physical universe in it was regarded as composed of discrete point-particles endowed with masses. They moved with time along well defined trajectories, in the fixed arena of a three-dimensional Euclidean space, under the influence of mutual forces. The trajectories could be deterministically calculated by using Newton's three laws of motion provided one knew the forces involved and also the initial position and velocities of all the particles. The forces involved were of the "action at a distance" type. Newton also discovered the universal attractive force of gravitation which acts between any two mass points and falls off as the square of the interparticle distance. Astronomy was thereby brought into the fold of physics unlike the case in Aristotlean physics of ancients.

It was known that there exists other forces such as magnetic forces, electric forces, chemical affinity, etc. It was part of post-Newtonian program of research to determine their laws. The force law between two "magnetic poles" was determined by John Mitchell in 1750, while that between two electric charges was conjectured theoretically by Joseph Priestley, the discoverer of Oxygen, in 1769 and experimentally verified in the unpublished work of Henry Cavendish done in 1771. It was however published first, based on his own work, by Charles Coulomb in 1785 and is now known as Coulomb's law. Alessandro Volta used electric currents, produced by his Voltaic pile, to dissociate a number of substances e.g. water into Hydrogen and Oxygen. After this work, it was a clear possibility that the forces responsible for chemical binding may be reducible to electrical forces. Matter could consist entirely of electrically charged mass points.

1.2. *Light as waves*

Newton was also inclined to view light also to be discrete stream of particles, 'light-corpuscles'. Christian Huygens communicated his researches on light to members of French Academy in 1678, and published in 1690 as 'Traité de la Lumiére', wherein he advanced the notion that light is a wave phenomena. The wave theory of light got strong boost from the discoveries of interference of light in 1801 by Thomas Young, and by the studies of Augustin Fresnel on diffraction of light begining in 1815. As a result, the wave theory of light was firmly established. It was inconcievable, in those days, to have a wave motion without a medium for it to propagate, so a "luminiferous aether" was postulated for its' propagation.

1.3. *Energetics program*

We just saw that light had proved refractory to being accomodated within Newton's mechanical conception of the universe. In thermodynamics, it was easy to see that the first law of thermodynamics, which refers to the law of energy conservation, could be easily interpreted within Newtonian framework. However it did not look possible to interpret the second law of thermodynamics, dealing with increasing entropy, within it. Ludwig Boltzmann's H-theorem was an attempt towards this goal during 1842–1877 using his kinetic theory of gases. This attempt attracted strong criticism from Ernst Zermelo and others. Georg Helm and Ludwig Ostwald, supported by Ernst Mach, therefore denied the reality of atoms and suggested that energy is the most fundamental concept and the whole program of physics should be reduced to a "generalized thermodynamics". This program, "Energetics", was subscribed to by a small but strongly vocal and influencial minority. In fact, Einstein's work on Brownian motion in 1905 played a crucial role in its fall.

1.4. *Electromagnetic conception of the universe*

Michael Faraday introduced the concept of continuous fields, like electric and magnetic fields, defined over the whole space-time, in contrast to discrete particles. He did this in order to have a deeper understanding of his law of electromagnetic induction in eighteen thirtees. These fields are produced by electric charges, and electric currents produced by these charges in motion. They then interact with other electric charges elsewhere. There is no "action at a distance" but every interaction is a local interaction.

Faraday quoted the old saying "matter cannot act where it is not" in a letter to Richard Taylor in 1844. Faraday also thought the gravitational force, which appears to act at a distance between two masses, could also be understood as a local interaction by the introduction of a gravitational field.

Clerk Maxwell's equations for electric and magnetic fields, given in 1864, unified these two disparate entities into a coherent single entity "electromagnetic field". Maxwell, synthesized the earlier known discoveries of Coulomb's law, Gauss' laws of magnetic induction, Oersted's work on production of magnetic fields by electric current, and Faraday's laws of electromagnetic induction into one set of equations using the field concept. He also appended a new element, now called "Maxwell's displacement current", to this synthesis.

A brilliant windfall from the Maxwell's equations was the prediction of the existence of transverse electromagnetic waves with a constant velocity (now denoted by the latter c). The velocity c agreed with the known velocity of light. It was therefore natural for Maxwell to propose "electromagnetic wave theory" of light. The subject of optics thus became a branch of electromagnetic theory. The luminiferrous aether was identified as the aether for electromagnetic fields as well.

The tantalizing possibility, the electromagnetic conception of the universe, arose now. Could it be that even point charged particles can be viewed as arising from the aether? The mass of an electron could be entirely due to its electromagnetic energy. If so, the "electromagnetic aether" would be the sole ontological entity in terms of which one would be able to understand the whole nature.

1.5. *Two clouds on the horizon*

In a lecture delivered in April 1900 before the Royal Institution, Lord Kelvin talked about two "Nineteenth Century Clouds Over the Dynamical Theory of Heat and Light". It was such a rare case of penetrating insight into the nature of physics that one is left admiring it even now. It is the resolution of these two "clouds" that gave rise to the two revolutions in twentieth century physics. One of these clouds referred to the continued unsuccessful attempts to detect the motion of the earth through aether and its resolution was achieved by Einstein's special theory of Relativity [1905]. We shall not be dealing with this any further here. The other cloud referred to the failure of the equipartition theorem in classical statistical mechanics. It resolution required the second revolution, associated with the quantum.

2. The Problem of Blackbody Radiation: From Kirchhoff to Planck

Max Planck, in 1900, was first to introduce the quantum ideas in physics and he did this in the context of blackbody radiation. We now discuss the early history of this problem for providing the setting of his work.

2.1. *Kirchhoff*

All heated bodies emit and absorb radiation energy. The emissivity $e(\lambda, T)$ of a body, for the radiation with wave length λ, depends on the nature of body and its temperature T. It is the same for its absorptivity $a(\lambda, T)$. Using the consideration of thermodynamics equilibrium, it was shown by Gustav Kichhoff of Berlin, in 1859, that the ratio of emissivity $e(\lambda, T)$ to its absorptivity $a(\lambda, T)$ is independent of the nature of the heated body i.e.

$$e(\lambda, T) = E(\lambda, T)a(\lambda, T)$$

where $E(\lambda, T)$ is a universal function of only the wave length λ of the radiation and its temperature T.

If we define, following Kirchhoff, a perfect blackbody as one whose absorptivity is equal to unity, i.e. perfect absorption, then the universal function $E(\lambda, T)$ can be identified with the emissivity of a perfect blackbody. He also showed that the radiation inside a heated cavity which is opaque and maintained at temperature T, behaves like blackbody radiation. One can therefore experimentally study the blackbody radiation by using the radiation issuing out a cavity through a small hole.

2.2. *Boltzmann*

Ludwig Boltzmann, in 1884, using Maxwell's electromagnetic theory showed that

$$E(\lambda, T) = (c/8\pi)\rho(\nu, T)\,,$$

where $\rho(\nu, T)$ is the energy density of radiation at frequency ν and temperature T. (c = velocity of light in vacuum, ν = frequency of the radiation $= c/\lambda$). He further showed using thermodynamics consideration, together with Maxwell's relation $P = \frac{1}{3}u$ between pressure P and energy density u of the radiation, that the total radiant energy per unit volution is proportional to T^4 i.e.

$$\int_0^\infty d\nu \rho(\nu, T) = \sigma T^4$$

where σ is called Stefan–Boltzmann Constant. Since Josef Stefan had conjectured the truth of this law on the basis of his experimental work in 1879 for all heated bodies, but it is strictly true only for a blackbody.

2.3. *Wien*

Further progress was made by Wilham Wien in 1894, when he studied the thermodynamics of extremely slow, i.e. adiabatic, contraction of the cavity on the blackbody radiation contained in it. From these, he concluded that

$$\rho(\nu, T) = \nu^3 f(\nu/T) .$$

This is known as 'Wien's displacement law'. We have thus reduced the problem of determining $\rho(\nu, T)$, a function of two variables ν and T, to that of determining a function $f(\nu/T)$ of a single variable (ν/T). This is as far as one can go on the basis of purely thermodynamic considerations.

To give a representation of the experimental data Wien also proposed a form for this function

$$\rho(\nu, T) = a\nu^3 e^{-b\nu/T} ,$$

which we shall refer to as Wien's radiation law. In this a and b are numerical coefficients to be fixed from the data.

2.4. *Rayleigh–Jeans*

In June 1900, Lord Rayleigh decided to apply equipartion theorem of Maxwell–Boltzmann to the problem of radiation and derived

$$\rho(\nu, T) = c_1 \nu^2 T .$$

He did not calculate at that time the numerical coefficient c_1, which he did in May 1905. He however, made a mistake of a factor of 8 which was corrected by James Jeans in June 1905. With the numerical factor included we have

$$\rho(\nu, T) = \frac{8\pi\nu^2}{c^3} \cdot kT$$

which is known as Rayleigh–Jeans' radiation law. Here k is the Boltzmann constant. Rayleigh felt that this is a limiting form of $\rho(\nu, T)$ for $\nu/T \to 0$.

Note that if this law was correct for all ν, then it would lead to ultraviolet catastrophe. The total energy would be infinite.

2.5. *Planck*

Max Planck succeeded to the chair of Kirchhoff at Berlin in 1889. He was naturally drawn to the problem of determining the universal function $\rho(\nu, T)$ introduced by his predecessor. As he said, "The so-called normal energy distribution represents something absolute, and since the search for absolutes has always appeared to me to be the highest form of research, I applied myself vigorously to its solution". He argued that since the universal $\rho(\nu, T)$ does not depend on the nature of the material of walls, its determination would be facilitated if one assumes a simple model for it. He proposed to regard the wall to be made of Hertzian oscillators, each one capable of emitting or absorbing radiation of only a single frequency ν. He then showed, using electromagnetic theory i.e.

$$\rho(\nu, T) = \frac{8\pi\nu^2}{c^3}\bar{E}(\nu, T)$$

where $\bar{E}(\nu, T)$ is the average energy of the Hertzian oscillator of frequency ν at temperature T. He had this result on May 18, 1899.

Earlier experimental work by Friedrich Paschen on blackbody radiation had shown that Wien's radiation law fitted the data well as it was known in 1897 for $\lambda = 1 - 8\mu$ and $T = 400 - 1600\ ^\circ K$. Later work by Otto Lummer and Ernst Pringhsheim, in the region $\lambda = 12 - 18\mu$ and $T = 300 - 1650\ ^\circ K$, had however revealed the deviations from Wien's radiation law in February 1900. On Oct 19, 1900 Kurlbaum announced the measurements done with Rubens for even higher wavelength region, $\lambda = 30 - 60\mu$ and $T = 200 - 1500\ ^\circ K$. Planck then gave his radiation law as a discussion remark to this announcement. In modern notation, (first done in 1906), it reads as

$$\rho(\nu, T) = \frac{8\pi\nu^2}{c^3} \cdot \frac{h\nu}{e^{h\nu/kT} - 1}$$

where h is now known as Planck's constant. This suggested radiation law fitted the data perfectly. Note also that it reduces to (i) Rayleigh–Jean's law for $\nu/T \to 0$ and (ii) has the same form as Wien's radiation law for $\nu/T \to \infty$ and (iii) provides the 'correct' interpolation formula between the two regions. At this stage it was a purely empirical formula without any derivation. He then got busy looking for one.

Planck, when he began his research career was inclined to the "energetics" school and believed in the deterministic significance, unlike what was advocated by Boltzmann who took the probabilistic view, of entropy. In Boltzmann's view the entropy S of a configuration was related to its thermodynamic probability W i.e.

$$S = k \ln W \, .$$

Planck, as an "act of desperation", was forced to use Boltzmann's view to derive his formula. In order to calculate thermodynamic probability for a configuration of N oscillators, with total energy $U_N = NU$ and entropy $S_N = NS$, he assumed that U_N is made up of finite energy elements ϵ i.e. $U_N = P\epsilon$, and worked out the total number of possible ways W_N of distributing P energy elements ϵ among N oscillators. He obtained

$$W_N = \frac{(N+P-1)!}{P!(N-1)!} \, .$$

The thermodynamic probability W was taken proportional to W_N. This leads to

$$S = \frac{S_N}{N} = k\left[\left(1+\frac{U}{\epsilon}\right)\ln\left(1+\frac{U}{\epsilon}\right) - \frac{U}{\epsilon}\ln\frac{U}{\epsilon}\right] .$$

On using $\frac{\partial S}{\partial U} = \frac{1}{T}$, we obtain

$$\bar{E}(\nu, T) = \frac{\epsilon}{e^{\epsilon/kT} - 1} \, ,$$

which on using Wien's displacement law, leads to (in modern notation)

$$\epsilon = h\nu \, .$$

Planck presented this derivation of his radiation law on 14 December 1900 to German Physical Society and this can be taken as the birth date of quantum theory. The really new element was his assumption that the Hertzian oscillators with frequency ν can emit or absorb radiation in the units of $\epsilon = h\nu$. Planck however did not realize the revolutionary nature of his procedure. As he said, "this was purely a formal assumption and I really did not give it much thought except that, no matter what the cost, I must bring about a positive result".

3. Einstein's Light Quantum Paper

3.1. *Light quantum hypothesis*

Albert Einstein was the first person to have a clear realization that Planck's introduction of energy quanta was a revolutionary step and thus one which would have larger significance for physics than just for the problem of black-body radiation. In 1905, Einstein's *annus mirabilis*, he published his light quantum paper.

Einstein started in this paper by first noting that the unambiguous prediction of electrodynamics and equipartition theorem for the material oscillators is that given by the radiation law, now called "Rayleigh–Jeans law". He was in fact the first person to derive this law from classical physics correctly as his work was done before Jeans obtained the proper numerical constant in it. As such Abram Pais, even felt that it would be more proper to call it Rayleigh–Einstein–Jean's law. Since this radiation law did not agree with experiments, and theoretically suffered from "ultraviolet catastrophe" (i.e. infinite total energy), it led to a clear failure of classical physics. Something in classical physics had to yield.

In his search for the cause of failure, Einstein was motivated by his dissatisfaction with asymmetrical treatment of matter and radiation in classical physics. As we saw earlier matter is discrete and particulate while the radiation is continuous and wave-field like in classical physics. He wondered whether the failure of the classical radiation theory was in not treating radiation also as discrete and particulate. He thus proposed his hypothesis of "light quantum". Of course he was well aware of the enormous success which wave theory of light had in dealing with the phenomenon of interference, diffraction, etc. of light. About this aspect, his comments:

"The wave theory, operating with continuous spatial functions, has proved to be correct in representing purely optical phenomena and will probably not be replaced by any other theory. One must, however, keep in mind that the optical observations are concerned with temporal mean values and not with instantaneous values, and it is possible, in spite of the complete experimental verification of the theory of reflection, refraction, diffraction, dispersion and so on that the theory of light which operates with continuous spatial functions may lead to contradictions with observations if we apply it to the phenomenon of generation and transformation of light".

Einstein then proceeded to show that an analysis of "experimental" Wien's radiation law, valid in "nonclassical" regime of large ν/T, gave an indication of the particle nature. For this purpose he did an elaborate calculation of the probability p that the monochromatic radiation of frequency ν, occupying a volume V_0, could all be found later in a volume V. He found it, on using Wien's radiation law, to be given by

$$p = (V/V_0)^n \text{ with } n = E/(h\nu),$$

(in modern notation), where E is the total energy. This is of the same form as that of a gas of n particles. From this remarkable similarity in the two results, he concluded "Monochromatic radiation of small energy density behaves, as long as Wien's radiation law is valid, for thermodynamic considerations, as if it consisted of mutually independent energy quanta of magnitude $R\beta\nu/N$". (The quantity $R\beta\nu/N$ is now denoted by $h\nu$.) This was the introduction by Einstein of light quanta hypothesis.

In the light quantum picture of Einstein "in the propagation of a light ray emitted from a point source, the energy is not distributed continuously over ever-increasing volumes of space, but consists of a finite number of energy quanta localized at points of space that move without dividing, and can be absorbed or generated as complete units". He then went on to apply the light quantum hypothesis to other phenomena involving the generation and transformation of light. The most important of these was his treatment of photoelectric effect. They also involved his successful application to eluciding the Stokes' rule in photoluminescence and to the ionization of a gas by ultraviolet light.

3.2. *The photoelectric effect*

In 1887, Heinrich Hertz observed that the ultraviolet light incident on metals can cause electric sparks. In 1899, J. J. Thomson established that the sparks are due to emission of the electrons. Phillip Lenard showed in 1902 that this phenomenon, now called the Photoelectric effect, showed "not the slightest dependence on the light intensity" even when it was varied to a thousandfold. He also made a qualitative observation that photoelectron energies increased with the increasing light frequency. The observations of Lenard were hard to explain on the basis of electromagnetic wave theory of light. The wave theory would predict an increase in photoelectron energy with increasing incident light intensity and no effect due to increase of frequency of incident light.

In Einstein's light quantum picture, a light quantum, with energy $h\nu$, when colliding with an electron in the metal, gives its entire energy to it. An electron from the interior of a metal has to do some work, W, to escape from the interior to the surface. We therefore get the Einstein photoelectric equation, for the energy of the electron E,

$$E = h\nu - W .$$

Of course, electron may lose some energy to other atoms before escaping to the surface, so this expression gives only the maximum of photo-electron energy which would be observed. One can see that Einstein's light quantum picture explains quite naturally the intensity independence of photoelectron energies and gives a precise quantitative prediction for its dependence on incident light frequency. It also predicts that no photoelectrons would be observed if $\nu < \nu_0$ where $h\nu_0 = W$. The effect of increasing light intensity should be an increase in the number of emitted electrons and not on their energy. Abram Pais called this equation as the second coming of the Planck's constant.

Robert A. Millikan spent some ten years testing Einstein equation and he did the most exacting experiments. He summarized his conclusions as well as his personal dislike of light quantum concept, as follows: "Einstein's photoelectric equation $\cdots$ appears in every case to predict exactly the observed results $\cdots$ yet the semi-corpuscular theory by which Einstein arrived at his equations seems at present wholly untenable" (1915) and "the bold, not to say reckless hypothesis of electromagnetic light corpuscle" (1916).

3.3. *Envoi*

Einstein's light quantum paper, which was titled, Über einen die Erzeugung und Verwandlung des Lichtes betreffenden heuristichen Geischtpunkt" (on a heuristic point of view concerning the generation and transformation of light), was completed on March 7, 1905 and appeared in *Annalen der Physik* **17**, 132–148 (1905) and was received by them on March 18, 1905.

It was thus his first paper during his *annus mirabilis* during which he also wrote papers on Brownian motion, special theory of relativity, and $E = mc^2$. Though, in public mind, he is associated indissolubly with relativity, with relativity as his most revolutionary contribution, Einstein himself regarded his light quantum paper among his papers written in 1905 as the "most revolutionary". The opinion of the recent historians of science tends to agree with Einstein. He was awarded the Nobel prize in physics in

1921 for this paper, and it was announced in Nov. 1922. Paranthetically his Nobel Lecture is on relativity theory.

Einstein's light-quantum is now known as "photon", a name given by G. N. Lewis, in as late as 1926. Though Einstein talked about photon energy $E = h\nu$, it is curious that he introduced the concept of photon momentum $\vec{p}$, with magnitude $|\vec{p}| = h\nu/c$ only in 1917. As we have seen, even Millikan did not believe in photon concept in 1915–1916 despite his having spent years on experimental work confirming it. In 1923, the kinematics of the Compton effect was worked out on the basis of it being an elastic electron-photon scattering by A. H. Compton. After that, it was generally accepted by physicists that light sometimes behaves as a photon.

4. Contributions to the Old Quantum Theory

4.1. *Specific heat of solids*

Both Planck in 1900, and Einstein 1905 used the quantum theory to understand problems of radiation. Einstein in 1907 was first to apply it to the problems of matter. This was the problem of specific heat of solids.

In 1819 Pierre Dulong and Alexis Petit, as a result of their joint experimental work on a number of metals and sulpher at room temperature, noted that all of them have almost the same specific heat C_V, at constant volume, with a value of 6 calories per mol. per $^\circ K$ i.e. $C_V = 3R$. Here R is universal gas constant. When other solids were investigated, especially carbon, the deviations were found from the Dulong–Petit Rule. In early 1870's Friedrich Weber conjectured and then verified that C_V approaches the value $3R$ even for those cases at higher temperature i.e. $C_V = 3R$ is only an asymptotic result. Theoretically Ludwig Boltzmann applied energy equipartition theorem to a three-dimensional lattice crystal and showed that $C_V = 3R$. However, the generality of the theorem left no scope for any deviations from this result within classical physics. There were similar problems which arose in the application of energy equipartition theorem for gases. As Lord Rayleigh noted in 1900 "What would appear to be wanted is some escape from the destructive simplicity of the general conclusions (following from energy equipartition theorem)". As we have noted earlier, Lord Kelvin regarded this problem as one of the clouds on the horizon of classical physics.

Einstein was first to realize that a use of equipartition theorem of classical statistics leads to Rayleigh–Jeans radiation law which was only asymptotically correct for large temperature. To get the correct Planck's radiation law one had to use quantum theory. It was therefore natural for him to try the same remedy to the problem of specific heat of solids. Besides he was always inclined to a symmetrical treatment of radiation and matter.

Einstein assumed a simple model of the solid. It is that of three-dimensional crystal lattice where all the atoms on the lattice oscillate harmonically and independently and with the same frequency. For a solid with N atoms we thus have a system of $3N$ harmonic oscillators of frequency ν. Using the earlier expression, in deriving Planck's expression for the average energy of an oscillator of frequency ν, and in thermal equilibrium at temperature T, we thus get for the total energy U of the solid,

$$U = 3N \cdot \frac{h\nu}{e^{h\nu/kT} - 1} .$$

This leads to Einstein's expression for specific heat for his model

$$C_V = 3R \frac{\xi^2 e^{\xi}}{(e^{\xi} - 1)^2} , \quad \xi = \frac{h\nu}{kT} .$$

It has the desirable feature that for ξ small i.e. large T, we get the Dulong–Petit result i.e.

$$C_V \longrightarrow 3R \text{ as } \xi \to 0 ,$$

which is the classical equipartion result. It provides a one parameter, i.e. ν, formula for the specific heat of a solid. The deviations from Dulong–Petit value are also in broad agreement with the experimental data. The model of solid assumed is too simplistic in that only a single frequency is assumed for all the oscillations. It was improved by Peter Debye in 1912, and a more exact treatment of atomic oscillations was given by Max Born and Theodore von Kármán in 1912–1913.

A preliminary formulation of the third law of thermodynamics was given by Walter Nernst in Dec. 1905 according to which the entropy of a system goes to zero at $T = 0$. Einstein's specific heat expression has the property that $C_V \to 0$ as $T \to 0$ and provides the first example of a model which is consistent with Nernst's heat theorem, as was noted by Nernst in 1910.

4.2. *Wave-particle duality*

In 1905, Einstein had used phenomenological Wien's radiation law to argue the particle nature of light. In 1909 he used Planck's radiation law to argue that light has both a particle and a wave aspect. For this purpose, he calculated an expression for the mean of square of energy fluctuations $\langle\epsilon^2(\nu,T)\rangle$ in the frequency interval ν and $\nu + d\nu$. From general thermodynamic considerations, we have

$$\langle\epsilon^2(\nu,T)\rangle = kT^2 v d\nu \frac{\partial\rho(\nu,T)}{\partial T},$$

in a subvolume v.

If we calculate this quantity using Rayleigh–Jeans radiation law $\rho = \rho_{R-J}(\nu,T)$, we obtain

$$\langle\epsilon^2(\nu,T)\rangle_{R-J} = \frac{e^2}{8\pi\nu^2}\rho^2_{R-J} v d\nu\,.$$

Note that Rayleigh–Jean derivation is based on the wave picture of light. If on the other hand, we calculate this quantity using Wien's radiation law, $\rho = \rho_{\text{Wien}}(\nu,T)$, we obtain

$$\langle\epsilon^2(\nu,T)\rangle_{\text{Wien}} = h\nu\rho_{\text{Wien}} v d\nu\,.$$

As we know, Wien's radiation law supports a particle picture of light.

We now use the correct Planck's law of radiation $\rho = \rho_{\text{Planck}}(\nu,T)$ and obtain

$$\langle\epsilon^2(\nu,T)\rangle_{\text{Planck}} = h\nu\rho_{\text{Planck}} v d\nu + \frac{c^2}{8\pi\nu^2}\rho^2_{\text{Planck}} v d\nu\,.$$

It is a very suggestive expression. The first term is of the form we obtain using Wien's law and supporting the particle picture light, while the second term has the same form as that given by Rayleigh–Jeans law which uses a wave picture of light. We also know that the contribution to the mean square fluctuations arising from independent causes are additive. This radiation has both wave and particle aspects. This was the first appearance in physics of wave-particle duality, here for light radiation.

Einstein was quite prophetic in his remarks on the implications of these results. He said "it is my opinion that the next phase in the development of theoretical physics will bring us a theory of light which can be interpreted as a kind of fusion of the wave and emission theory ⋯ wave structure and quantum structure ⋯ are not to be considered as mutually incompatible ⋯. We will have to modify our current theories, not to abandon them completely".

4.3. *Einstein's A and B coefficients and the discovery of stimulated emission of light*

In 1916–1917 Einstein gave a new and wonderful derivation of Planck's radiation law which provides a lot of new insights. As he wrote to his friend Michel Besso, in 1916, "A splendid light has dawned on me about the absorption and emission of radiation".

He considered the thermodynamic equilibrium of a system comprising a gas of "molecules" and radiation. The "molecules" here referred to any material system which is interacting with radiation. Let the energy levels of the "molecules" be denoted by E_m and let the number of "molecules" be given by N_m when they occupy the energy level E_m.

Consider two of these levels E_2 and E_1 with $E_2 > E_1$ and consider the transitions from level 2 to level 1 and the reverse. Einstein postulated that the number of transitions, in time dt, in the "molecules" for the higher state E_2 to the lower state E_1 consists of two components. One of these due to spontaneous jumps from E_2 to E_1. The number of transitions however is given by the term $A_{21}N_2dt$. Here the coefficient A_{21} is related to the intrinsic probability of this jump and does not depend on the radiation density. The second of these is due to stimulated emission of radiation. The number of transitions is here taken to be given by the term $B_{21}N_2\rho dt$ and is taken proportional to the radiation density ρ. Here the coefficient $B-21$ is related to the probability of this process. The presence of radiation will also induce transitions from the lower level 1 to higher level 2. The number of these transitions is taken to $B_{12}N_1\rho dt$ and is again taken proportional to the radiation density ρ. The coefficient B_{12} again is related to the probability of this process. The A_{ij}'s and B_{ij}'s are called Einstein's A and B coefficients.

In equilibrium, the number of transitions from level 1 to level 2 must be same as the number of transitions from level 2 to level 1. We therefore get the relation

$$N_2(A_{21} + B_{21}\rho) = N_1 B_{12}\rho\,,$$

or

$$\rho = \frac{(A_{21}/B_{21})}{\left(\frac{B_{12}}{B_{21}}\right)\left(\frac{N_2}{N_1}\right) - 1}\,.$$

Following Boltzmann, we have

$$N_m = p_m e^{-E_m/kT}\,,$$

where p_m is the relevant weight factor, and using it, we get

$$\rho = \frac{(A_{21}/B_{21})}{\left(\frac{B_{12}p_2}{B_{21}p_1}\right) e^{(E_1-E_2)/kT} - 1} .$$

From Wiens displacement we conclude that

$$E_2 - E_1 = h\nu ,$$

a relation given by Bohr in 1913. These transitions must involve emission or absorption of radiation of frequency ν. Further for large temperatures, i.e. $T \to \infty$, the ρ must reduce to Rayleigh–Jean's law. This is possible only if we have

$$\frac{A_{21}}{B_{21}} = \frac{8\pi h\nu^3}{c^3}$$

$$p_2 B_{12} = p_1 B_{21} .$$

Through this analysis we have got insights into the probabilities of transitions and correct quantitative relations between them. A calculation of these was not possible until the full apparatus of quantum electrodynamics was in place which came much later, only in 1927.

The concept of stimulated emission, given by the coefficient B_{21}, was introduced by Einstein here for the first time. He was forced to take this step, since otherwise he would have been led to Wien's radiation law by these considerations and not to the correct Planck's law. This concept is of fundamental importance in the theory of lasers.

5. Quantum Statistics: Bose and Einstein

The last great contribution to quantum theory, before the advent of quantum mechanics, by Einstein was to develop quantum statistics for a system of material particles. Here, the original idea was due to the Indian physicist — Satyendranath Bose from Dacca University — and was given in the context of radiation theory. Einstein extended it to matter. As such this quantum statistical method is known as Bose Statistics or Bose–Einstein statistics. All integral spin particles in nature have been found to obey this statistics and are called "Bosons". All half-odd integral spin particles obey Fermi–Dirac statistics, which was given later in 1926 and are called "Fermions".

5.1. *Bose*

On June 4, 1924 Bose sent a short paper to Einstein containing a new derivation of Planck's law. It was accompanied by a very unusual request to translate it into German and get it published in *Zeitschrift für Physik*, if he found it worthwhile. Bose explained his chutzpah in doing it by saying "Though a complete stranger to you, I do not feel any hesitation in making such a request, because we are all your pupils though profiting only by your teachings through your writings". He also mentioned that he "was the one who translated your paper on Generalized Relativity" when the first ever English translation of the relativity papers of Einstein was published by the Calcutta University in 1920. We also know now, through William Blanpied, that this paper had earlier been rejected for publication by the *Philosophical Magazine*.

Bose noted "since it's (Planck's law's) publication in 1901, many methods for deriving this law have been proposed $\cdots$. In all cases it appears to me that the derivations have not been sufficiently justified from a logical point of view. As opposed to these, the light quantum combined with statistical mechanics (as formulated to meet the needs of the quantum) appears sufficient for the derivation of the law independent of the classical theory".

Bose's idea was to regard the blackbody radiation as a free photon gas and then treat it by the method of statistical mechanics. This was his strategy to derive Planck's radiation law in a logically consistent manner.

Now photons of frequency ν have energy $h\nu$ and a momentum, with magnitude $ph\nu/c$, on the light quantum hypothesis of Einstein. A straightforward calculation of the phase space volume element leads to the factor $4\pi p^2 dpV$, where V is the volume of the gas. Bose multiplied it by a further factor of 2, in order to take into account the two polarization states of the light, to obtain $8\pi p^2 dpV$. If we now divide it by a factor h^3, following Planck's proposal of 1913 "that phase space cells have a volume h^3" we obtain for the number of phase space cells in this phase space volume element $8\pi p^2 dpV/h^3$. This leads to, using $p = h\nu/c$, the first factor $8\pi\nu^2 d\nu/c^3$ in the Planck's radiation law. Bose has thus shown that the number A^S of the phase space cells between radiation frequency ν^s and $\nu^s + d\nu^s$ to be given by

$$A^S = \frac{8\pi(\nu^s)^2 V d\nu^s}{c^3}$$

in a novel way. Note that Bose obtained this factor here, unlike Planck, without making any use of the electromagnetic theory. Bose emphasized this aspect of his derivation in his letter to Einstein.

If Bose had proceeded further and used the statistical methods of Boltzmann, at this stage, he would have obtained Wien's law and not the desired Planck's law. He however chose to interpret A^S, not as the number of "particles" but as number of "cells", which played the role of "particles" in Boltzmann's counting. This procedure then led to Planck's law. This is equivalent to treating photons as indistinguishable in contrast to classical Boltzmann statistics where particles are identical but distinguishable. To give a simple example, if we have to distribute two identical balls, which are distinguishable, by being colored red and blue, into three containers, there are nine possible different configurations and probability of each one is 1/9 (Boltzmann counting). On the other hand, if two identical balls are not distinguishable, as we are color blind, then there are only six possible different configurations. This is so since the red ball in one container and blue ball in the other container are indistinguishable from the configuration in which we interchange the two balls. The probability of each distinct configuration flow is now 1/6 (Bose counting).

5.2. *Einstein*

Einstein immediately saw the importance of Bose's work and got it published in *Zeitschrift für Physik* after translating it into German together with an appreciative note. Not only that, in view of his predilection to treat radiation and matter on the same footing, he extented it immediately to a gas of material particles during 1924–1925. For a photon gas there is no constraint of holding the total number of photons fixed but for material particles, let us say "atoms", we have also a new constraint to hold the total number fixed. This introduced another parameter, chemical potential, which has to be determined using this constraint. Bose did not comment on the indistinguishably aspect in his paper. To bring this aspect out Einstein also rewrote the Bose's formula for the total number of configuration in the form it is normally found in textbooks.

We have seen that Einstein's model of solids was the first known example in which Nernst's theorem was valid. The case of Bose–Einstein gas, which Einstein worked out, provides the first model of a gas for which Nernst's theorem holds.

Einstein also studied the fluctuations for the ideal Bose–Einstein gas, as he had done earlier for radiation. On calculating the mean square fluctuation (Δn^2) for the number $n(\epsilon)$ of atoms having energy between ϵ and $\epsilon + d\epsilon$, he found it to consist again of two terms

$$(\Delta n)^2 = n(\epsilon) + \frac{n^2(\epsilon)}{Z(\epsilon)}$$

where $Z(\epsilon)$ is the number of particle states in the energy interval ϵ and $\epsilon + d\epsilon$. The first term is the expected one for particles.

For an interpretation of the second term, which implies a wave aspect for matter, Einstein suggested that this is due to the wave nature of atoms as postulated by Louis de Broglie in his recent doctoral thesis of 1924. Einstein was aware of this thesis as Pierre Langevin had sent him a copy for his opinion, and it was only Einstein's favorable comments on it which made Langevin accept de Broglie's thesis. Einstein also suggested associating a scalar field with these waves.

5.3. *Bose–Einstein condensation*

A free boson gas undergoes a phase transition below a critical temperature T_{BE}. A macroscopic fraction of the atoms condense into the lowest energy state. This phase transition is not due to interparticle attractive interaction but is simply a manifestation of the tendency of bosons to stick together. This was again a first solvable model for a phase transition.

Despite a lot of efforts it was not possible to experimentally test this prediction of Bose–Einstein until quite late. It was finally observed only in 1995. The Nobel Prize in Physics for the year 2001 was awarded to Eric Cornell, Carl Wieman and Wolfgang Ketterle for this discovery.

6. Foundations of Quantum Mechanics

6.1. *Discovery of quantum mechanics*

After a quarter century of long and fruitful interaction between the old quantum theory and the experimental work on atomic systems and radiation, this heroic period came to an end in 1925 with the discovery of Quantum Mechanics. It was discovered in two different mathematical formulations viz first as Matrix Mechanics and a little later as Wave Mechanics.

Werner Heisenberg discovered Matrix mechanics during April–June 1925. A complete formulation was achieved by Max Born, Werner Heisenberg and Pascual Jordan in October 1925. After the mathematical formalism was in place, the problems of its interpretation arose. At Copenhagen, Niels Bohr and Heisenberg and others devoted their full attention to this talk. The resulting interpretation, called 'The Copenhagen Interpretation of Quantum Mechanics', was to dominate the physics, despite some other contenders, for a long time. Heisenberg proposed his famous 'uncertainty principle' in Feb. 1927 in this connection. In this work he was strongly influenced by a conversation he had with Einstein in 1926 at Berlin. Heisenberg acknowledged to Einstein the role which relativity with its analysis of physical observation had played in his own discovery of matrix mechanics. His motivation in formulating it had been to rid the theory of physical unobservables. Einstein differed and said "it is nonsense even if I had said so ⋯ on principle it is quite wrong to try founding a theory on observables alone ⋯. It is the theory which decides what is observable".

The second formulation, wave mechanics, was published during the first half of 1926, as a series of four papers "Quantization as an Eigenvalue problem" in *Annalen der Physik* by Erwin Schrödinger. He was led to study the papers of de Broglie, wherein he suggested that matter should also exhibit a wave nature, through a study of Einstein's papers on Bose–Einstein gas. He prefered a wave theory treatment to the photon treatment of Bose and could avoid new statistics. As he said "That means nothing else but taking seriously the de-Broglie–Einstein wave theory of moving particles" in a paper on Bose–Einstein gas theory. His next step was to make the idea of matter-waves more precise by writing a wave equation for them. This is the famous Schrödinger wave equation for matter waves resulting in the birth of wave mechanics. As Schrödinger acknowledged "I have recently shown that the Einstein gas theory can be founded on the consideration of standing waves which obey the dispersion law of de Broglie ⋯. The above considerations about the atom could have been presented as a generalization of these considerations". As Pais says "Thus Einstein was not only one of three fathers of the quantum theory but also the sole godfather of wave mechanics". The three fathers alluded to here are Planck, Einstein and Bohr.

The mathematical equivalence of these two formulations was soon established by Schrödinger and Carl Eckart in 1927.

After the discovery of quantum mechanics the focus of Einstein shifted from applications of quantum theory to various physical phenomena to the problems of understanding what the new mechanics means. With his deep committment to the reality of an objective world Einstein was not in tune with the Copenhagen interpretation.

6.2. Discussions at Solvay conferences

The fifth Solvay Conference was held at Brussels in October 1927. It was in this meeting that the claim of completeness of quantum mechanics as a physical theory was put forward first. In this connection Einstein discussed the example of single hole diffraction of the electron in order to illustrate two contrasting points of view:

(i) "the de Broglie–Schrödinger waves do not correspond to a single electron but to a cloud of electrons extended in space. The theory does not give any information about the individual processes", and
(ii) "the theory has the presentations to be a complete theory of individual processes".

The first viewpoint is what is now known as statistical or ensemble interpretation of quantum mechanics if we clarify the phrase "a cloud of electrons" to refer to an ensemble of single electron systems rather that to a many electron system. This is the view which Einstein held in his later work. He was thus the originator of "The Statistical or Ensemble interpretation of Quantum Mechanics". This view was also subscribed to by many others including Karl Popper and Blokhintsev. It is essentially the minimalist interpretation of quantum mechanics.

The second view point is the one upheld by the Copenhagen School and very many others and may be termed as the maximalist interpretation. Here a pure state provides the fullest description of an individual system e.g. an electron.

The setup envisaged by Einstein was as follows: Consider a small hole in an opaque screen and let an electron beam fall on it from the left side. Let it be surrounded by another screen, on the right side, a hemispherical photographic plate. From quantum mechanics the probability of an electron hitting at any point of the photographic is uniform. In the actual experiment, the electron will be found to have been recorded at a single definite point on the plate. As Einstein noted that one has to "presuppose a very peculiar mechanism of action at a distance which would prevent the wave function, continuously distributed over space from acting at two places of the screen simultaneously $\cdots$ if one works exclusively with Schrödinger waves, the second interpretation of ψ in my opinion implies a contradiction with the relativity principle". Here Einstein is worried about, what we now call "the collapse of the wave function" postulate and its consistency with special theory of relativity. Einstein therefore opted for the statistical interpretation of quantum mechanics. A detailed discussion of this interpretation would be out of place here.

Apart from the formal discussion remark of Einstein noted above there were also lots of informal discussions between him and Niels Bohr. In these discussions Einstein generally tried to evade or violate Heisenberg's uncertainty relations for individual processes by imagining various possible experimental setups and Bohr constantly tried to find the reason as why they would not work. The uncertainties involved were taken to be due to errors involved in the simultaneous measurement of position-momentum or energy-time pairs. These discussion continued also at Solvay Conference held in 1930. These dialogues are quite famous and Niels Bohr wrote an elegant account of them later. It is generally agreed that in these discussions Bohr was successful in convincing Einstein that it was not possible to evade the uncertainty principle. However later developments, such as Bohm's realistic model have shown that these discussions are somewhat irrelevant to the problem of interpretation of quantum mechanics.

6.3. *Quantum nonseparability and Einstein–Podolsky–Rosen correlations*

In quantum mechanics, if two systems have once interacted together and later separated, no matter how far, they cannot any more be assigned separate state vectors. Since physical interaction between two very distant systems is negligible, this situation is very counterintuitive. Schrödinger even emphasized this aspect, "I would not call that one but rather the characteristic of quantum mechanics". More technically, this is so for all two particle systems having a nonseparable wave function. A wave function is regarded as nonseparable, if no matter what choice of basis for single particle wave function is used, it cannot be written as a product of single particle wave functions. Such wave functions are called entangled. The entanglement is a generic feature of two particle wave functions.

In 1935, A. Einstein, B. Podolsky and B. Rosen (EPR) published a paper "Can Quantum Mechanical Description of Reality be Considered Complete?" in *Physical Review*. It had a rather unusual title for a paper for this journal. In view of this they provided the following two definitions at the beginning of the paper:

(1) A *necessary* condition for the *completeness* of a theory is that every element of the physical reality must have a counterpart in the physical theory.

(2) A *sufficient* condition to identify an element of reality: "If, without in any way disturbing a system, we can predict with certainty (i.e. with probability equal to unity) the value of a physical quantity, then there exists an element of physical reality corresponding to this physical quantity".

We now illustrate the use of these definitions for a single-particle system. Let the position and momentum observable of the particle be denoted by Q and P, respectively. Since in an eigenstate of Q, we can predict with certainty the value of Q, which is given by its eigenvalue in that eigenstate, it follows that the position Q of the particle is an element of physical reality (e.p.r.). Similarly the momentum P is also an e.p.r. The position Q and the momentum P however are not simultaneous e.p.r. So at the single particle level there is no problem with quantum mechanics, as far as these definitions of 'completeness' and 'elements of reality' are concerned.

Interesting new things are, however, encountered when a two-particle system is considered. Let the momenta and position of the two particles be denoted respectively by P_1 and Q_1 for the first particle and by P_2 and Q_2 for the second particle. Consider now the two-particle system in the eigenstate of the relative-position operator, $Q_2 - Q_1$ with eigenvalue q_0. The relative position $Q_2 - Q_1$ can be predicted to have a value q_0 with probability one in this state and thus qualifies to be an e.p.r. We can also consider an eigenstate of the total momentum operator, $P_1 + P_2$, with an eigenvalue p_0. The total momentum can be predicted to have a value p_0 with probability one and thus also qualifies to be an e.p.r. Furthermore relative position operator, $Q_2 - Q_1$, and total momentum operator, $P_1 + P_2$, commute with each other and thus can have a common eigenstate, and thus qualify to be *simultaneous* elements of physical reality.

We consider the two-particle system in which two particles are flying apart from each other having momenta in opposite directions and are thus having a large spatial separation. The separation will be taken so that no physical signal can reach between them. Let a measurement of position be made on the first particle in the region R_1 and let the result be q_1. It follows from standard quantum mechanics that instantaneously the particle 2, which is a spatially far away region R_2, would be in an eigenstate $q_0 + q_1$ of Q_2. The Q_2 is thus an e.p.r., the position of second particle gets fixed to the value $q_0 + q_1$, despite the fact that no signal can reach from region R_1 to R_2 where the second particle is, a "spooky action at a distance" indeed. On the other hand, a measurement of the momentum P_1 of the first particle, in the region R_1 can be carried out and let it result in a measured value p_1. It then follows from the standard quantum mechanics, that the particle 2, in the region R_2 would be in an eigenstate of its momentum P_2 with an eigenvalue $p_0 - p_1$. The p_2 is thus also an e.p.r. This however leads to a contradiction since Q_2 and P_2 cannot be a simultaneous e.p.r. as they do not commute. We quote the resulting conclusion following from this argument as given by Einstein in 1949.

EPR Theorem: The following two assertions are not compatible with each other

(1) the description by means of the ψ-function is complete
(2) the real states of spatially separated objects are independent of each other.

The predilection of Einstein was that the second postulate, now referred to as "Einstein locality" postulate, was true and thus EPR theorem establishes the incompleteness of quantum mechanics.

As Einstein said "But on one supposition we should in my opinion, absolutely hold fast: the real factual situation of the system S_2 is independent of what is done, with system S_1, which is spatially separated from the former".

Einstein, Podolsky and Rosen were aware of a way out of the above theorem but they rejected it as unreasonable. As they said "Indeed one would not arrive at our conclusion if one insisted that two or more quantities can be regarded as simultaneous elements of reality only when they can be simulateneously measured or predicted. On this point of view, either one or the other, but not both simultaneously, of the quantities P and Q can be predicted, they are not simultaneously real. This makes the reality of P and Q depend upon the process of measurement carried out on the first system, which does not disturb the second system in any way. No reasonable definition of reality could be expected to permit this".

6.4. *Later developments*

David Bohm reformulated the Einstein–Podolsky–Rosen discussion in a much simpler form in terms of two spin one-half particles in a singlet state in 1951. This reformulation was very useful to John Bell, who in 1964, gave his now famous Bell-inequalities on spin correlation coefficients following from Einstein locality for EPR correlations. These inequalities are experimentally testable. In experiments of increasingly higher precision and sophistication, they have shown agreement with quantum mechanics and a violation of local realism though some loopholes remain. Bell's work on hidden variable theories and Einstein–Podolsky–Rosen correlations had a profound influence on the field of foundations of quantum mechanics, in that it moved it from a world of sterile philosophical discussions to a world of laboratory experiments.

More recently E.P.R. correlations and quantum entanglement has been found useful in developing new technologies of quantum information such as quantum cryptography, quantum teleportation. They have ceased to be embarrasments but are seen as useful resources provided by quantum mechanics. There are even hopes of developing quantum computing which would be much more powerful that the usual universal Turing machines.

Einstein's legacy in physics still looms large. Talking about his work, Max Born once said "In my opinion he would be one of the greatest theoretical physicists of all times even if he had not written a single line of relativity".

Bibilographical Notes

A brief and nontechnical summary of this paper appeared in Singh, V. [2005], The Quantum Leap, *Frontline*, **22**, #10, 22–24. It also overlaps in places with author's earlier related writings cited in various notes below.

References

[1] On the life and science of Einstein, the literature is enormous. The best biography for physicists is,
Pais, A. [1982] '*Subtle is the Lord* ··· ': *The Science and the Life of Albert Einstein* (Clarendon Press, Oxford: Oxford University Press, New York).
It however needs supplementing by more recent work on foundational aspects of quantum mechanics: Also important is
Schilpp, P. A. (ed.) [1949] *Albert Einstein: Philospher Scientist*, Library of Living Philosophers Vol. 7, La Salle, Ill.: Open Court (Harper Torch Books reprint, 1959).
It contains Einstein's "Autobiographical notes" and "Reply to criticism" as well as a "Bibilography of the writings of Albert Einstein to May 1951" apart from 25 papers on the science and philosophy of Einstein by various authors.
At a more popular level, we have
French, A. P. (ed.) [1979] *Einstein: A Centenary Volume* (Harvard University

Press, Cambridge, MA);
Bernstein, J. [1973] *Einstein* (The Viking Press, New York).

[2] (a) For the writings of the Einstein we have the multivolume ongoing series, [1987] *The Collected Papers of Albert Einstein* (Princeton University Press, Princeton, NJ),
and the companion volumes
[1987] *The Collected Papers of Albert Einstein: English Translation* (Princeton University Press, Princeton, NJ).
His papers from the miraculous year 1905 are available in English translation also in
Stachel, J. (ed.) [1998] *Einstein's Miraculous Year: Five Papers that Changed the Face of Physics* (Princeton) (Indian reprint: Scientia, An imprint of Srishti Publishers for Centre for Philosophy and Foundations of Science, New Delhi, 2001).
(b) for the references to some specific Einstein papers discussed in this paper, see
(i) [1905] *Light quantum paper*: *Annalen der Physik*, **17**, 132–148.
(ii) [1907] *Specific heat of solids*: *Annalen der Physik*, **22**, 180–190, 800.
(iii) [1909] *wave-particle duality*: *Phys. Zeitschrift*, **10**, 185, 817.
(iv) [1916] *A- and B-coefficients*: *Verh. Deutsch Phys. Ges.* **18**, 318; [1916] *Mitt. Phy. Ges. (Zurich)*, **16**, 47; [1917] *Phys. Zeitschrift* 18, 121.
The concept of photon momentum is introduced in the 1917 paper mentioned here.
(v) [1924] *Bose–Einstein Gas*: *Sitzungber. Preus. Akad. Wiss. Phys. Math. Kl.*, p. 261; [1925] p. 3 and [1928] p. 18.
(vi) *Einstein-Bohr dialogues*: Bohr, N., Discussions with Einstein on Epistemological Problems in Atomic Physics, in the *Schilpp Volume* cited earlier.
(vii) *E.P.R. Theorem*: Einstein, A., Podolsky, B. and Rosen, N. [1935] *Phys. Rev.* **57**, 777, and *Schilpp Volume* cited earlier.
(c) Other English translations of his light quantum paper and 1917 paper on A-B coefficients are also available in
Boorse, H. A. and Motz, L. (eds.) [1966] *The World of Atoms* (Basic Books, New York); and
ter Haar, D. [1967] *The Old Quantum Theory* (Pergamon, Oxford).
An English translation of Einstein's first two paper on Bose–Einstein statistics is available also in
Sengupta, N. D. [1983] *Phys. News* **14**, 10 and [1983] **14**, 36.
Einstein quotes used in the text are from various sources and are sometimes slightly modified or abridged.

[3] For Physics before Einstein, see
Whittaker, E. T. [1960] *A Theory of Aether and Electricity, Vol. 1, Classical Theories* (Harper Torchbacks);
Bork, A. M. [1966] *Science* **152**, 597;
Singh, V. [1980] *Science Today*, p. 19–23.

[4] For a historical account of Quantum Mechanics, see
Whittaker, E. T. [1960] *A Theory of Aether and Electricity, Vol. 2, Modern*

Theories (1900–1926) (Harper Torchbacks);
Jammer, M. [1966] *The Conceptual Development of Quantum Mechanics* (New York);
Hermann, A. [1971] *The Genesis of Quantum Theory (1899–1913)*, translated by Nash, C. W. (MIT Press, Cambridge, MA),
Hund, F. [1974] *The History of Quantum Theory*, translated by G. Reece (Harrap, London);
Kragh, H. [2001] *Quantum Generations: A History of Physics in the Twentieth Century* (Princeton Univ. Press) (Indian reprint, Universities Press, Hyderabad, 2001).

[5] The work of S. N. Bose appeared in
Bose, S. N. [1924] *Zeits. fur Physik* **26**, 178 and [1924] **27**, 384.
Two English translations exist of both the papers, one by
Banerjee, B. [1974] *Phys. News* **5**, 2, 40, 42
and another by
Theimer, O. and Ram, B. [1976] *Am. J. Phys.* **44**, 1058 and [1977] **45**, 242.
About his life and science, see
Blanpied, W. A. [1972] *Am. J. Phys.* **40**, 1212;
Singh, V. [1974] *Science Today*, p. 29–34;
Mehra, J. [1975] *Biographical Memories of the Fellows of Royal Society, London* **21**, 117;
Chatterjee, S. D. [1983] *Biographical Memories of the Fellows of Indian National Science Academy*, 7, 59.

[6] For later developments on the foundations of quantum mechanics see,
Ballentine, L. E. [1970] *Rev. Mod. Phys.* **42**, 358 and [1972] *Am. J. Phys.* **40**, 1763;
Jammer, M. [1974] *The Philosophy of Quantum Mechanics* (New York);
Bell, J. [1987] *Speakable and Unspeakable in Quantum Mechanics* (Cambridge);
Selleri, F. [1988] *Quantum Mechanics versus Local Realism: The Einstein–Podolsky–Rosen Paradox* (Plenum);
Home, D. [1997] *Conceptual Foundations of Quantum Physics* (Plenum, New York).
Nielsen, M. and Chuang, I. L. [2000] *Quantum Computation and Quantum Information* (Camabridge);
Singh, V. [2004] Quantum mechanics and reality, arXive: quant-ph/0412148;
Singh, V. [2005] Hidden variables, noncontextuality and Einstein locality in quantum mechanics, arXive: quant-ph/0507182.

CHAPTER 10

Einstein's Legacy: Relativistic Cosmology

****

JAYANT V. NARLIKAR

Inter-University Centre for Astronomy and Astrophysics, Post Bag 4, Ganeshkhind, Pune 411 007, India

This review gives a historical account of how cosmology has developed since the 1917 paper of Albert Einstein. Current frontier level science draws on contemporary astronomy as well as contemporary physics, stretching both as far as extrapolations will permit. Thanks to numerous observations at different wavelengths, cosmologists today have their plates full. Extrapolations of laboratory tested physics are required for to understand all information within the framework of a standard model. The success and shortcomings of this approach are briefly discussed against the historical backdrop.

1. Historical Background

Two years after proposing his general theory of relativity in 1915, Albert Einstein [1] used it in an ambitious way to propose a model of the entire universe. This simple model assumed that the universe is homogeneous and isotropic and *also static*. Homogeneity means that the large scale view of the universe and its physical properties at any given epoch would be the same at all spatial locations. Isotropy demands that the universe looks the same in all directions, when viewed from any spatial location. The requirement of a static universe was motivated by the perception then that there is no large-scale systematic movement in the universe.

That was the general belief at the time. In fact the realization that there is a vast world of galaxies spread beyond the Milky Way had not yet seeped into the astronomical community. Although there were isolated measurements of nebular redshifts, these did not convey any impression that the universe as a whole is not static. However, to obtain such a static model Einstein had to modify his general relativistic field equations to include an additional *cosmological constant term* λ which corresponded to a long range force of repulsion.

The original equations were:

$$R_{ik} - 1/2 g_{ik} R = -[8\pi G/c^4] T_{ik} \,. \tag{1}$$

Here the left-hand side relates to the spacetime geometry of the universe and the right-hand side describes the physical contents of the universe. These equations did not yield a static solution and so Einstein sought to modify them in the *simplest possible way.* This led him to the following equations:

$$R_{ik} - 1/2 g_{ik} R + \lambda g_{ik} = -[8\pi G/c^4] T_{ik} \,. \tag{2}$$

In the "Newtonian approximation" this additional term corresponds to an acceleration of $\lambda r c^2$ between any two matter particles separated by a distance r. The constant λ is called the cosmological constant since its value is very small (today's estimate is $\sim 10^{-56}$ cm^{-2}) and it does not affect the motion of matter significantly on any but the cosmological scale.

The *Einstein Universe,* as the model came to be known described the universe by a spacetime metric given by

$$ds^2 = c^2 dt^2 - S^2[dr^2/(1-r^2) + r^2(d\theta^2 + \sin^2\theta d\phi^2)] \,, \tag{3}$$

where the spherical polar coordinates have their usual meaning on the surface of a hypersphere of radius S. The field Eq. (2) then gives the density and radius of the universe in terms of the fundamental constants G, c and λ. To Einstein this was an eminently satisfactory outcome as it related physics of the universe to its spacetime geometry in a unique way. The gravity of the matter "curled up" the space into a finite volume, showing the essence of the general relativistic relationship between gravity and space curvature. He felt that the uniqueness of the solution attached special significance to the model in terms of credibility.

He was in for disappointment on this count as within a few months de Sitter [2] found another solution to the same equations with the metric given by

$$ds^2 = c^2 dt^2 - e^{2Ht}[dr^2 + r^2(d\theta^2 + \sin^2\theta d\phi^2)] \,, \tag{4}$$

where H = constant. The *de Sitter Universe* was homogeneous and isotropic but *non-static.* It described an expanding but empty universe.

One can say that whereas the Einstein universe had matter without motion, the de Sitter universe had motion without matter. In 1917, the astronomical data did not support the de Sitter model, which remained a mathematical curiosity.

In 1922–1924, Alexander Friedmann [3], however, showed that one can obtain homogeneous and isotropic solutions without the cosmological term, but they describe models of an *expanding universe*. In 1927, Abbé Lemaître [4] also obtained similar solutions, but these, along with the Friedmann models were considered as mathematical curiosities.

Meanwhile, on the observational side, the early (pre-1920) perception of a universe mostly confined to the Milky Way Galaxy with the Sun at its center, eventually gave way to the present extra-galactic universe in which our location has no special significance. Indeed this 1905 quotation of Agnes Clerke [5] in her popular book on astronomy expresses the current dogma of those times:

> *The question whether nebulae are external galaxies hardly any longer needs discussion. It has been answered by the progress of research. No competent thinker, with the whole of the available evidence before him, can now, it is safe to say, maintain any single nebula to be a star system of co-ordinate rank with the Milky Way. A practical certainty has been attained that the entire contents, stellar and nebula, of the sphere belong to one mighty aggregation, and stand in ordered mutual relations within the limits of one all embracing scheme.*

This perception represented the majority view which was still current in 1920 when the famous Shepley–Curtis debate [6] took place. Shapley spoke in support of this view while Curtis represented the slowly emerging view that many of the faint nebulae were external galaxies far away from the Milky Way.

During the 1920s Edwin Hubble gradually established this picture in which spiral and elliptical galaxies are found all over the universe. The erroneous observations of Van Maanen [7] contradicting this picture and arguing that all spiral nebulae were galactic, had been influential in the delay in accepting this revised picture. These were eventually set aside. In 1929, Hubble established what is today known as the Hubble Law [8] which is generally interpreted as coming from an expanding universe. In this Hubble spectroscopically determined the Doppler radial velocities of galaxies and found these to vary in proportion to their distances. The constant of proportionality is called the Hubble constant and today it is denoted by H. Thus one may write Hubble's law in terms of redshifts as:

$$z = (H/c) \cdot D\,, \tag{5}$$

where D is the distance of the extragalactic object with redshift z. The Friedmann–Lemaître models now no longer were mathematical curiosities but were seen as the correct models to explain Hubble's law. They were all describable with the line element

$$ds^2 = c^2dt^2 - S^2[dr^2/(1 - kr^2) + r^2(d\theta^2 + \sin^2\theta d\phi^2)]\,, \tag{6}$$

where the parameter k takes values 1, 0 or -1. The Einstein universe had $k = 1$ whereas the de Sitter universe had $k = 0$. The coordinates r, θ, ϕ are constant for a typical galaxy and may be called its comoving coordinates. The motion of the galaxy is manifest through the scale factor $S(t)$. The redshift is interpreted in terms of this model as coming from a time-dependent increasing scale-factor $S(t)$: if the light signal from the source left at time t_1 and it reached the observer at time t_0 then we have

$$1 + z = S(t_0)/S(t_1)\,. \tag{7}$$

The scale-factor $S(t)$ and the curvature parameter k were to be determined from Einstein's field equations. Einstein also decided that his cosmological constant was no longer needed and gave it up. Incidentally the much-publicised remark by Einstein that the cosmological constant was the "greatest blunder" of his life has no direct authentication in Einstein-literature. It has been ascribed to George Gamow who claimed that this is what Einstein said to him [9].

The stage was thus set to launch cosmology as a discipline wherein the theoretical predictions based on relativistic models could be tested by observations of the extragalactic universe.

2. Early Cosmology

During the 1930s, cosmologists led by Eddington [10] and Lemaître [11] discussed the theoretical models of the expanding universe and all these led to the concept of a "beginning" when the universe was dense and very violent. Lemaître called the state that of a *primeval atom*. Later, Fred Hoyle, an opponent of this idea referred to the state as of "big bang", a name that caught on when the model became more popular.

The crucial effect in Hubble's law was the redshift found in the spectra of galaxies and its progressive increase with the galactic distances. The linear law discovered by Hubble was believed to be an approximation of the exact functional relationship between redshift and distance according to any of the various Friedmann–Lemaître models. Attempts were made by succeeding astronomers to carry out deeper surveys to test the validity of this extrapolation. This will be discussed later.

Hubble's own priorities on the observational side, were elsewhere [12]. He wanted to fix the value of the mathematical parameter k of the model by observing galaxies and counting them to larger and larger distances. He made several unsuccessful attempts before realizing that the ability of the 100-inch Hooker telescope fell short of making a significant test of the relativistic models. The 5-metre telescope at the Palomar Mountain was proposed by him for this very reason as this bigger telescope was expected to settle this cosmological problem. By the time the telescope was completed and began to function (late 1940s) Hubble had realized that his observational programme was not a realistic one and the telescope in fact came to be used for other important works.

The reason Hubble's programme was unworkable was that in order to detect the effects of spacetime curvature through galaxy counts, one needed to look very far, out to redshifts of the order unity, and this requirement was hard to satisfy for two reasons. (1) Observational techniques were not yet sophisticated enough to detect galaxies of such large redshifts. (2) The number of galaxies to be counted was enormously large if one were to use the counts to be sensitive enough to draw cosmological conclusions. There was a third difficulty with the number count programme, to which I shall return in Sec. 9.

3. The Advent of Radio Astronomy

Astronomy became more versatile after World War II, after radio astronomy came into existence as a viable tool of observations. In their enthusiasm about the new technique, radio astronomers felt that they could undertake Hubble's abandoned programme by applying it to the counts of radio sources. In the 1950s radio astronomers in Cambridge, England and in Sydney as well as Parkes, Australia, began their attempts to solve this problem by counting radio sources out to very faint limits. Radio astronomy apparently got round the two difficulties mentioned above. Radio galaxies could be observed, it was felt, to greater distances than optical galaxies and there were far fewer of them to count.

The basic test of counting of radio sources went thus. If one accepts that radio sources are of uniform luminosity and are homogeneously distributed in the universe, then in the static Euclidean model, it can be easily shown that the number (N)–flux density (P) relation satisfies the relation

$$\log N = -1.5 \log P + \text{constant}\,. \tag{8}$$

The relation for a typical expanding Friedmann–Lemaître model shows a relation starting with Eq. (8) at high flux end and getting flatter at low fluxes. If, however, one put in an *ad hoc* assumption that the number density of radio sources per unit comoving coordinate volume was higher than at present, then one could get slopes steeper than -1.5.

While the Australians felt that within the existing error-bars, their surveys did not show any evidence inconsistent with the Euclidean model, the Cambridge group under the leadership of Martin Ryle made several claims to have found a steep slope. While the early Cambridge data were later discounted as being of dubious accuracy, the data in the early 1960s (the 3C and 4C surveys) did show a slope of -1.8 at high flux density, which subsequently flattened at low flux densities. The steepness was claimed by Ryle to have confirmed the big bang models. However, it later became clear that these radio surveys might tell us more about (1) local inhomogeneity and (2) the physical properties of the sources rather than about large scale geometry of the universe [13].

4. The Steady State Theory

In 1948, there emerged a rival to the classic big bang theory. Authored by Hermann Bondi, and Thomas Gold [14] and independently by Fred Hoyle [15], this theory was based on a model of the universe with the de Sitter metric, but which had a constant non-zero density of matter. Such a model can be obtained from Einstein's gravitational equations (without the cosmological term), provided on the right-hand side one introduces a negative energy field, called originally the C-field. Hoyle and later Maurice Price (private communication) worked on the C-field concept and a theory based on a scalar field derivable from an action principle emerged in 1960. This idea was developed further by Hoyle and Narlikar [16]. Although the concept of a negative energy scalar field was considered by physicists to be unrealistic in the 1960s, today, four decades later it is appreciated that the currently popular phantom fields are no different from the C-field.

Since, as the name implies, the steady state theory described an unchanging universe (on a large enough scale), the observational predictions of the theory were unambiguous and this was cited as a strength of the theory. Ryle's main attack was directed against this theory with the assertion that the radio source counts disproved this theory. This claim was refuted by Hoyle and Narlikar [17] with the demonstration that in a more realistic structure of the universe inhomogeneities on the scale of 50–100 Mpc (megaparsec: 1 parsec is approximately 3 light years) would give rise to steep slopes of the $\log N$–$\log P$ curve for radio sources.

Although the steady state theory survived Ryle's challenges, it appeared to receive a mortal blow in 1965 by the discovery of the cosmic microwave background. Also, it could not account for the rather large fraction ($\sim$25%) by mass of helium in the universe. To understand the implications of this result one needs to look back at the studies of the early universe in relativistic cosmology.

5. The Early Hot Universe

In the mid-1940s, George Gamow [18, 19] started a new programme of studying the physics of the big bang universe close to the big bang epoch. For example, calculations showed that the universe in its early epochs was dominated by relativistically moving matter and radiation and that the temperature T of the universe, infinite at the big bang, dropped according to the law:

$$T = B/S \cdot B = \text{constant}\,. \tag{9}$$

Thus it fell to about ten thousand million degrees after one second. In the era 1–200 second, Gamow expected thermonuclear reactions to play a major role in bringing about a synthesis of the free neutrons and protons that were lying all over the universe. Were all the chemical elements we see today in the universe formed in this era?

This expectation of Gamow turned out to be incorrect. Only light nuclei, mainly helium could have formed this way. Also, one could adjust the density of matter in the universe over a wide band to produce the right cosmic abundance of helium. The heavier elements could, however, be formed in stars, as was shown later by the comprehensive work of Geoffrey and Margaret Burbidge, William Fowler and Fred Hoyle [20]. Today it looks as if the light nuclei were made in Gamow's early universe, as the stars do not seem to be able to produce them in the right abundance. It was because of this circumstance that the steady state universe which did not have a very hot era, failed in the production of helium.

Apart from this evidence, there was another prediction [21] made by Gamow's younger colleagues, Ralph Alpher and Robert Herman, namely that the radiation surviving from that early hot era should be seen today as a smooth Planckian background of temperature of around 5 K. This prediction has been substantiated. In fact in 1941, McKeller [22] had deduced the existence of such a background of temperature 2.3 K from spectroscopic observations of CN and other molecules in the galaxy. This result was not widely known or appreciated at the time. In fact it was the serendipitous observation of an isotropic radiation background in 1965 by Arno Penzias and Robert Wilson [23] that drew physicists and cosmologists to the big bang model in a big way. Penzias and Wilson found the temperature to be $\sim$3.5 K.

The post-1965 development of cosmology took a different turn. The finding of the cosmic microwave background radiation (CMBR) was taken as vindication of the early hot universe and on the observational side efforts were made to observe the spectrum of the radiation as accurately as possible. In 1990, the COBE satellite gave a very accurate Planckian spectrum [24] thus providing confirmation of the Alpher–Herman expectation of a relic black body spectrum. Another expectation, of finding small scale inhomogeneities in the background was also fulfilled two years later when COBE found [25] such fluctuations of temperatures $\Delta T/T$ of the order of a few parts in a million. On the theoretical side the emphasis shifted from general relativistic models to models of a very small scale universe with high temperature corresponding to fast moving particles. Theorists also began to come to grips with the problem of formation of large-scale structure ranging from galaxies to superclusters. We will consider these developments next.

6. Physics of the Early and Very Early Universe

The cosmic microwave background radiation (CMBR) prompted many physicists to look in depth at the physics of the post- and pre-nucleosynthesis era. For example, as the universe cools down, the chemical binding can become important and trap the free electrons into protons to make neutral hydrogen atoms. This eliminates the major scattering agency from the universe and radiation can subsequently travel freely. Calculations [26] show that this epoch was at redshift of around 1000–1100.

If instead we explore epochs *earlier* than the nucleosynthesis one, we would encounter larger temperature and more energetic activity. This has attracted particle physicists to the big bang models for here they have a possibility of testing their very high energy physics. The very early epochs when the universe was 10^{-38} second old had particles of energy so high that they might have been subject to the grand unification scheme which could therefore be tested. Energies required for such testing are, however, some 13 orders of magnitude higher than what can be produced by the most powerful accelerators on the Earth.

Such a combination of disciplines is called *astroparticle physics*. One of its most influential 'gifts' has been the notion of inflation [27]. This is the rapid exponential expansion of the universe lasting for a very short time, produced by the phase transition that took place when the grand unified interaction split into its component interactions (the strong and electroweak interactions). Inflation is believed to solve some of the outstanding problems of the standard big bang cosmology, such as the horizon problem, the flatness problem, the entropy problem, etc. Another article by Sarkar in this volume deals with the main aspects of astroparticle physics.

7. Dark Matter and Dark Energy

One of the conclusions of inflation is that the space part of the universe is flat. Theoretically it requires the matter density to be $\rho_c = 3H^2/8\pi G$. Here H is the Hubble constant and G is the gravitational constant. This value, sometimes known as the *closure density*, leads straightaway to a conflict with primordial nucleosynthesis which tells us that at this density there would be almost no deuterium produced. Even if we ignore inflation, and simply concentrate on the empirical value of matter density determined by observations, we still might run into a serious conflict between theory and observation: there is evidence for greater matter density than permitted by the above deuterium constraint.

For, while the visible matter in the form of galaxies and intergalactic medium leads to a value of density which is less than 4% of the closure density, there are strong indications that additional *dark* matter may be present too [13]. The adjective 'dark' indicates the fact that this matter is unseen but exerts gravitational attraction on visible matter. Such evidence is found in the motions of neutral hydrogen clouds around spiral galaxies and in the motions of galaxies in clusters. Even this excess matter would cause problem with deuterium.

To get round this difficulty, the big bang cosmologists have hypothesized that the bulk of dark matter is *non-baryonic*, that is it does not influence nucleosynthesis. Writing the ratio of the density of non-baryonic matter to the closure density as Ω_{nb} and the corresponding ratio for baryonic matter as Ω_b, we should get as per inflation $\Omega_{nb} + \Omega_b = 1$. Thus if the baryonic matter is 4%, the non-baryonic matter should be 96%.

However, even this idea runs into difficulty as there is no direct evidence for so much dark matter. A solution is provided, however, by resurrecting the cosmological constant that Einstein had abandoned in the 1930s. We can define its relative contribution to the dynamics of expansion through a parameter analogous to the density parameter:

$$\Omega_\Lambda = 3\lambda H^2/c^2 \,. \tag{10}$$

Thus we now get something like: $\Omega_b = 0.04$, $\Omega_{nb} = 0.23$, and $\Omega_\Lambda = 0.73$. This extra energy put in is called *dark energy*. The total of these values is meant to add up to unity, as expected by the inflationary hypothesis.

8. Structure Formation

These issues are important to the understanding of how large scale structure developed in the universe. To this end, the present attempts assume that small fluctuations were present in the very early universe and these grew because of inflation and subsequent gravitational clustering. Various algorithms exist for developing this scenario. One of the basic inputs is the way the total density is split up between baryonic matter, non-baryonic matter and dark energy. The non-baryonic dark matter can be hot (HDM) or cold (CDM) depending on whether it was moving relativistically or non-relativistically at the time it decoupled from ordinary (baryonic) matter.

A constraint to be satisfied by this scenario is to reproduce the observed disturbances found in the CMBR by these agents and also the observed extent of clustering of galaxies today. For, observations of small inhomogeneities of the CMBR rule out various combinations and also suggest what kind of dark matter (cold or hot or mixed) might be required. Currently the model favored is called the ΛCDM-model to indicate that it has dark energy and cold dark matter.

9. Observational Tests

Like any physical theory cosmology also must rely on observational tests and constraints. There are several of these. There have been tests of cosmological models of the following kinds: (i) Geometry of the universe; (ii) Physics of the universe.

The first category includes the measurement of Hubble's constant, the redshift magnitude relation to high redshifts, the counting of radio sources and galaxies, the variation of angular size with redshift and the variation of surface brightness with redshift. The measurement of Hubble's constant has been a tricky exercise right from the early days dating back to Hubble's original work. The problem is to be sure that no systematic errors have crept in the distance measurement, as these have not yet been fully debugged. Which is why we still have serious observing programmes yielding values close to 70 km/s/Mpc as well as to 55 km/s/Mpc. At the time of writing this review, the majority opinion favors the higher value but 'rule of the majority' has not always been a successful criterion in cosmology.

The measurement of z–m relation had been attempted by Allan Sandage for quite a long time and during the period 1960–1990 the overall view was that the relation as applied to brightest galaxies in clusters treated as standard candles, favored *decelerating* models. These models are naturally given by the Friedmann solutions *without* the cosmological constant. However, in the late 1990s, the use of Type Ia supernovae has led to a major reversal of perception and the current belief [28] is that the universe is *accelerating*. The other tests like number counts or angular size variation have not been so clearcut in their verdict as they get mixed up with evolutionary parameters. Apart from the difficulties encountered by Hubble in the 1930s, any cosmological test using source populations of a certain type necessarily gets involved with the possibility that the source yardstick may be evolving with age.

Currently cosmologists are most attracted to measurements of the angular power spectrum of the microwave background inhomogeneities. These can be related to other dynamical features of the universe, given a cosmological model satisfying Einstein's equations with the cosmological constant. Using the details from WMAP satellite [29] one can get a range of models with $k = 0$. Among these models those with a positive cosmological constant are favored. As mentioned before, the favored solution has $\Omega_b = 0.04$, $\Omega_{nb} = 0.23$, and $\Omega_\Lambda = 0.73$. We recall that the low value of baryonic density is required to understand the abundance of deuterium.

Many cosmologists feel that there is now a 'concordance' between various tests that suggest the above combination for the energy content of the universe together with the higher of the two values of the Hubble constant mentioned above. It is felt that this set of parameters describes accurately most of the observed features of the universe. With this optimistic view one may be tempted to think that the quest for the model of the universe that began with Einstein in 1917 is coming to an end.

10. Need for Caution and Alternatives

However, there needs to be some caution towards this optimism. The concordance has been achieved at the expense of bringing in a lot of speculative element into cosmology. Thus there is as yet no independent evidence for the non-baryonic dark matter, nor any for the dark energy. When one finds that these two make up more than 96% of matter in the universe leaving only about 4% to the astronomer for direct observation, one wonders whether the claims based on the unseen and the untested are really as firm as one wants in science. Then a lot revolves round the concept of inflation which is still not describable as a process based on a firm physical theory. Nor is the inflationary era observable by any telescopes today. The densities of matter one is talking about when inflation took place were some 10^{50} times the density of water. Recall how much investigation went into the equation of state for neutron stars where the matter density was a mere 10^{15} times the density of water. Yet one finds no discussions of such esoteric matter amongst the cosmologists. Likewise, the inflationary time scales of the order of 10^{-38} second defy any operational physical meaning. These are some twenty five orders of magnitude smaller than the shortest measurable time scale known to physics, viz. those measured by the atomic clocks. So a physicist may wonder if the concordence cosmology is a rigorous physical exercise at all.

The concordance picture looks good today if one is happy with the number of epicycles that have gone into it. Non-baryonic dark matter and dark energy are two of them. They had to be introduced in order to ensure the survival of the model: they have no independent direct confirmation.

These are examples of extrapolations of known physics to epochs that are astronomically unobservable. While indirect observations showing an overall consistency of these assumptions are necessary for the viability of the concordance model, they cannot be considered sufficient.

This is why there appears to be the need for new ideas in cosmology especially alternative scenarios that are less speculative and follow very different tracks from the above standard scenario. Some attempts are in vogue at present, like the Quasi-Steady State Cosmology (QSSC) [30] or the Modified Newtonian Dynamics (MOND) [31], which are, however very much minority efforts. Perhaps by 2017, a hundred years after Einstein's paper on cosmology we may have a more realistic perception of how complex our universe is. I can do no better than end with a quotation from Fred Hoyle [32]:

> '... *I think it is very unlikely that a creature evolving on this planet, the human being, is likely to possess a brain that is fully capable of understanding physics in its totality. I think this is inherently improbable in the first place, but even if it should be so, it is surely wildly improbable that this situation should just have been reached in the year 1970 ...* '

Fred Hoyle said this at the Vatican Conference held towards the end of the 1960–1970 decade when cosmologists were making equally confident remarks about how well the universe was being understood. This was before inflation, dark matter, dark energy, etc. were even thought of. Are today's cosmologists sure that they have all pieces of the jigsaw puzzle that make up our universe?

References

[1] Einstein, A. [1917] *Preuss. Akad. Wiss. Berlin Sitzber.*, 142.
[2] de Sitter, W. [1917] *Koninkl. Akad. Weteusch Amsterdam* **19**, 1217.
[3] Friedmann, A. [1922] *Z. Phys.* **10**, 377; [1924] *Z. Phys.* **21**, 326.
[4] Lemaître, A. G. [1927] *Annales de la Société Scientifique de Bruxelles* **XLVIIA**, 49.
[5] Clerke, A. [1905] *The System of the Stars* (Adam and Charles Black, London), p. 349.
[6] Shapley, H. and Curtis, H. D. [1921] *Bull. Nat. Res. Council* **2**, part 3, no. 11.
[7] van Maanen, A. [1916–1930] *Mt Wilson Contr.* Nos. 111, 136, 158, 182, 204, 237, 270, 290, 321, 356, 391, 405–408.

[8] Hubble, E. P. [1929] *Proc. Natl. Acad. USA* **15**, 168.
[9] Gamow, G. [1971] *My World Line* (Viking Adult, USA).
[10] Eddington, A. S. [1930] *Mon. Not. R. Astron. Soc.* **90**, 668.
[11] Lemaître, A. G. [1931] *Mon. Not. R. Astron. Soc.* **91**, 490.
[12] Hubble, E. P. [1938] *Astrophys. J.* **84**, 517.
[13] Narlikar, J. V. [2002] *An Introduction to Cosmology* (Cambridge), 3rd edn.
[14] Bondi, H. and Gold, T. [1948] *Mon. Not. R. Astron. Soc.* **108**, 252.
[15] Hoyle, F. [1948] *Mon. Not. R. Astron. Soc.* **108**, 372.
[16] Hoyle, F. and Narlikar, J. V. [1962] *Proc. R. Soc.* **A270**, 334; [1966] *Proc. R. Soc.* **A290**, 162.
[17] Hoyle, F. and Narlikar, J. V. [1961] *Mon. Not. R. Astron. Soc.* **123**, 133.
[18] Gamow, G. [1946] *Phys. Rev.* **70**, 572.
[19] Alpher, R. A., Bethe, H. A. and Gamow, G. [1948] *Phys. Rev.* **73**, 80, because of the names of the authors this work is sometimes referred to as the $\alpha\beta\gamma$ theory!
[20] Burbidge, E. M., Burbidge, G. R., Fowler, W. A. and Hoyle, F. [1957] *Rev. Mod. Phys.* **29**, 547. This work is referred to as the B^2FH theory.
[21] Alpher, R. A. and Herman, R. C. [1948] *Nature* **162**, 774.
[22] McKellar, A. [1941] *Pub. Dom. Astrophys. Obs. Victoria* **7**, 251.
[23] Penzias, A. A. and Wilson, R. W. [1965] *Astrophys. J.* **142**, 419.
[24] Mather, J. C. *et al.* [1990] *Astrophys. J.* **354**, L37.
[25] Smoot, G. *et al.* [1992] *Astrophys. J.* **396**, L1.
[26] Weinberg, S. [1972] *Gravitation and Cosmology* (John Wiley).
[27] Kazanas, D. [1980] *Astrophys. J.* **241**, L59; Guth, A. H. [1981] *Phys. Rev. D* **23**, 347; Sato, K. [1981] *Mon. Not. R. Astron. Soc.* **195**, 467: all independently suggested the idea, but Guth's version and terminology caught on.
[28] Reiss, A. *et al.* [1998] *Astrophys. J.* **116**, 1009; Perlmutter, S. *et al.* [1999] *Astrophys. J.* **517**, 565.
[29] Spergel, D. N. *et al.* [2003] *Astrophys. J. Suppl.* **148**, 175.
[30] Hoyle, F., Burbidge, G. and Narlikar, J. V. [2000] *A Different Approach to Cosmology* (Cambridge University Press, Cambridge).
[31] Milgrom, M. [1983] *Astrophys. J.* **271**, 365, 371 and 383.
[32] Hoyle, F. [1970] in *Study Week on Nuclei of Galaxies*, ed. O'Connell (North Holland, Amsterdam), p. 757.

CHAPTER 11

Einstein's Universe: The Challenge of Dark Energy

****

SUBIR SARKAR
Rudolf Peierls Centre for Theoretical Physics,
1 Keble Road, Oxford OX1 3NP, UK
s.sarkar@physics.ox.ac.uk

From an observational perspective, cosmology is today in excellent shape — we can now look back to when the first galaxies formed ~ 1 Gyr after the Big Bang, and reconstruct the thermal history back to the primordial nucleosynthesis era when the universe was $\sim 1\,\mathrm{s}$ old. However, recent deep studies of the Hubble diagram of Type Ia supernovae indicate that the expansion rate is accelerating, requiring the dominant component of the universe to have *negative* pressure like vacuum energy. This has been indirectly substantiated through detailed studies of angular anisotropies in the cosmic microwave background and of spatial correlations of the large-scale structure of galaxies, which also require most of the matter component to be non-baryonic. Although there are plausible candidates for the constituent of the dark matter in physics beyond the Standard Model (e.g. supersymmetry), the energy scale of the required 'dark energy' is $\sim 10^{-12}\,\mathrm{GeV}$, well below any known scale of fundamental physics. This has refocussed attention on the notorious cosmological constant problem at the interface of general relativity and quantum field theory. It is likely that the resolution of this issue will require fundamental modifications to Einstein's ideas about gravity.

1. Introduction

In the accompanying article, Jayant Narlikar recounts how in 1917 Einstein boldly applied his newly developed theory of general relativity to the universe as a whole [1]. The first cosmological model was *static* to match prevalent ideas about the universe (which, at that time, was confined to the Milky way!) and to achieve this, Einstein introduced the 'cosmological constant' term in his equation [2]. Within a decade it had become clear from the work of Slipher and Hubble that the nebulae on the sky are in fact other 'island universes' like the Milky Way and that they are receeding from us — the universe is expanding. Einstein wrote to Weyl in 1933: "*If there is no quasi-static world, then away with the cosmological term*".

This however was not easy — after all the symmetry properties of Einstein's equation do allow any constant proportional to the metric to be added to the left-hand side:

$$R_{\mu\nu} - \frac{1}{2} g_{\mu\nu} R + \lambda g_{\mu\nu} = \frac{-8\pi T_{\mu\nu}}{M_{\mathrm{P}}^2}, \tag{1}$$

where we have written Newton's constant, $G_{\mathrm{N}} \equiv 1/M_{\mathrm{P}}^2$ (where $M_{\mathrm{P}} \simeq 1.2 \times 10^{19}\,\mathrm{GeV}$), in natural units ($\hbar = k_{\mathrm{B}} = c = 1$). Moreover, with the subsequent development of quantum field theory it became clear that the energy-momentum tensor on the right-hand side can also be freely scaled by another additive constant proportional to the metric which reflects the (Lorentz invariant) energy density of the vacuum:

$$\langle T_{\mu\nu} \rangle_{\mathrm{fields}} = -\langle \rho \rangle_{\mathrm{fields}} g_{\mu\nu} \,. \tag{2}$$

This contribution from the matter sector adds to the "bare" term from the geometry, yielding an effective cosmological constant:

$$\Lambda = \lambda + \frac{8\pi \langle \rho \rangle_{\mathrm{fields}}}{M_{\mathrm{P}}^2}, \tag{3}$$

or, correspondingly, an effective vacuum energy: $\rho_{\mathrm{v}} \equiv \Lambda M_{\mathrm{P}}^2 / 8\pi$.

For an (assumed) homogeneous and isotropic universe with the Robertson–Walker metric $\mathrm{d}s^2 \equiv g_{\mu\nu} \mathrm{d}x^\mu \mathrm{d}x^\nu = \mathrm{d}t^2 - R^2(t)[\mathrm{d}r^2/(1-kr^2) + r^2 \mathrm{d}\Omega^2]$, we obtain the Friedmann equations describing the evolution of the cosmological scale-factor $R(t)$:

$$H^2 \equiv \left(\frac{\dot{R}}{R}\right)^2 = \frac{8\pi}{3M_{\mathrm{P}}^2}\rho - \frac{k}{R^2} + \frac{\Lambda}{3}, \qquad \frac{\ddot{R}}{R} = -\frac{4\pi}{3M_{\mathrm{P}}^2}(\rho + 3p) + \frac{\Lambda}{3}, \tag{4}$$

where $k = 0, \pm 1$ is the 3-space curvature signature and we have used for "ordinary" matter (and radiation): $T_{\mu\nu} = p g_{\mu\nu} + (p+\rho) u_\mu u_\nu$, with $u_\mu \equiv \mathrm{d}x_\mu / \mathrm{d}s$. The conservation equation $T^{\mu\nu}_{;\nu} = 0$ implies that $\mathrm{d}(\rho R^3)/\mathrm{d}R = -3pR^2$, so that given the 'equation of state parameter' $w \equiv p/\rho$, the evolution history can be constructed. Since the redshift is $z \equiv R/R_0 - 1$, for non-relativistic particles with $w \simeq 0$, $\rho_{\mathrm{NR}} \propto (1+z)^{-3}$, while for relativistic particles with $w = 1/3$, $\rho_{\mathrm{R}} \propto (1+z)^{-4}$, but for the cosmological constant, $w = -1$ and ρ_v =constant! Thus radiation was dynamically important only in the early universe (in fact for $z \gtrsim 10^4$) and for most of the expansion history only non-relativistic matter is relevant. The Hubble equation can then be rewritten with reference to the present epoch (subscript 0) as

$$H^2 = H_0^2[\Omega_{\rm m}(1+z)^3 + \Omega_k(1+z)^2 + \Omega_\Lambda]\,,$$
$$\Omega_{\rm m} \equiv \frac{\rho_{\rm m0}}{\rho_{\rm c}}, \quad \Omega_k \equiv -\frac{k}{R_0^2 H_0^2}, \quad \Omega_\Lambda \equiv \frac{\Lambda}{3H_0^2}\,, \tag{5}$$

yielding the sum rule $\Omega_{\rm m} + \Omega_k + \Omega_\Lambda = 1$. Here $\rho_{\rm c} \equiv 3H_0^2 M_{\rm P}^2/8\pi \simeq (3 \times 10^{-12}\,{\rm GeV})^4 h^2$ is the 'critical density' for a $k = 0$ universe (in the absence of Λ) and the present Hubble parameter is $H_0 \equiv 100h$ km s^{-1} Mpc^{-1} with $h \simeq 0.7$, i.e. about 10^{-42} GeV.

As Weinberg discussed in his influential review in 1989 [3], given that the density parameters $\Omega_{\rm m}$ and Ω_k were observationally known to be not much larger than unity, the two terms in Eq. (3) are required to somehow conspire to cancel each other in order to satisfy the approximate constraint

$$|\Lambda| \lesssim H_0^2, \tag{6}$$

thus bounding the present vacuum energy density by $\rho_{\rm v} \lesssim 10^{-47}\,{\rm GeV}^4$ which is a factor of 10^{123} below its 'natural' value of $\sim M_{\rm P}^4$ — the cosmological constant problem! The major development in recent years has been the recognition that this inequality is in fact saturated with $\Omega_\Lambda \simeq 0.7$, $\Omega_{\rm m} \simeq 0.3$ ($\Omega_k \simeq 0$) [5], i.e. there *is* non-zero vacuum energy of just the right order of magnitude so as to be detectable today.

In the Lagrangian of the Standard Model of electroweak and strong interactions, the term corresponding to the cosmological constant is one of the two 'super-renormalizable' terms allowed by the gauge symmetries, the second one being the quadratic divergence in the mass of fundamental scalar fields due to radiative corrections [4]. To tame the latter sufficiently in order to explain the experimental success of the Standard Model has required the introduction of a supersymmetry between bosonic and fermionic fields which is (softly) broken at about the Fermi scale. Thus the cutoff scale of the Standard Model, viewed as an effective field theory, can be lowered from the Planck scale $M_{\rm P}$ down to the Fermi scale, $M_{\rm EW} \sim G_{\rm F}^{-1/2} \sim 300$ GeV, albeit at the expense of introducing about 150 new parameters in the Lagrangian, as well as requiring delicate control of the many non-renormalizable operators which can generate flavor-changing neutral currents, nucleon decay, etc., so as not to violate experimental bounds. This implies a *minimum* contribution to the vacuum energy density from quantum fluctuations of $\mathcal{O}(M_{\rm EW}^4)$, i.e. "halfway" on a logarithmic scale down from the Planck scale to the energy scale of $\mathcal{O}(M_{\rm EW}^2/M_{\rm P})$ corresponding to the observationally indicated vacuum energy. Thus even the introduction of supersymmetry cannot eradicate a discrepancy by a factor of at least 10^{60} between "theory" and observation.

It is generally believed that a true resolution of the cosmological constant problem can only be achieved in a full quantum theory of gravity. Recent developments in string theory and the possibility that there exist new dimensions in Nature have generated many interesting ideas concerning possible values of the cosmological constant [2, 6]. Nevertheless it is still the case that there is *no* generally accepted solution to the enormous discrepancy discussed above. Of course the cosmological constant problem is not new but there has always been the expectation that somehow we would understand one day why it is exactly zero. However, if it is in fact non-zero and dynamically important *today*, the crisis is much more severe since it also raises a cosmic 'coincidence' problem, viz. why is the present epoch special? It has been suggested that the 'dark energy' may not be a cosmological *constant* but rather the slowly evolving potential energy $V(\phi)$ of a hypothetical scalar field ϕ named 'quintessence' which can *track* the matter energy density. This however is also fine-tuned since one needs $V^{1/4} \sim 10^{-12}\,\text{GeV}$ but $\sqrt{\text{d}^2V/\text{d}\phi^2} \sim H_0 \sim 10^{-42}\,\text{GeV}$ (in order that the evolution of ϕ be sufficiently slowed by the Hubble expansion), and moreover, does not address the fundamental issue, viz. why are all the other possible contributions to the vacuum energy absent? Admittedly the latter criticism also applies to attempts to do away with dark energy by interpreting the data in terms of modified cosmological models. Given the 'no-go' theorem against dynamical cancellation mechanisms in Eq. (3) in the framework of general relativity [3], it might appear that solving the problem will necessarily require our understanding of gravity to be modified. However, to date no such alternative which is phenomenologically satisfactory has been presented. The situation is so desperate that 'anthropic' arguments have been advanced to explain why the cosmological constant is just of the right order of magnitude to allow of our existence today, notwithstanding the fact that we have little or no understanding of its prior probability distribution!

Given this sorry situation on the theoretical front, this article will focus solely on the new observational developments and present a critical assessment of the evidence for dark energy. It is no exaggeration to say that this tiny energy density of the present vacuum poses the biggest challenge that fundamental theory and cosmology have ever faced.

2. The Observational Situation

That we live in an universe which has evolved from a hot dense past is now well established, primarily on the basis of the exquisite match (see Fig. 1) of a Planck spectrum to the intensity of the cosmic microwave background (CMB), with

$$T_0 = 2.725 \pm 0.001\text{K}, \tag{7}$$

as determined by the *COBE* satellite [7]. It has also been shown that the blackbody temperature does increase with the redshift as $T = T_0(1 + z)$, by observing fine-structure transitions between atomic levels of C I in cold gas clouds along the line of sight to distant quasars [10]. These observations are difficult to accomodate within the 'Quasi-Steady State Cosmology' in which the CMB arises through thermalization of starlight [11] — a mechanism that was already severely constrained by the closeness of the observed CMB spectrum to the Planck form [12]. Moreover, the absence of spectral distortions requires that the evolution must have been very close to adiabatic (i.e. constant entropy per baryon) back at least to the epoch when the universe was dense enough and hot enough for radiative photon creation processes to be in equilibrium (at $z \gtrsim 10^7$) [8].

A modest extrapolation in redshift back to $z \sim 10^{10}$ takes us back to the epoch of Big Bang nucleosynthesis (BBN) when the weak interactions interconverting neutrons and protons became too slow to maintain chemical equilibrium in the cooling universe, and the subsequent nuclear reactions rapidly converted about 25% of the total mass into the most stable light nucleus ^{4}He. Trace amounts of D, ^{3}He and ^{7}Li were also left behind with abundances sensitive to the baryon density. As shown in Fig. 2, the primordial abundances of all these elements as inferred from a range of observations are in reasonable agreement with the standard calculation [9]. Moreover, the implied baryon-to-photon ratio $\eta = n_\text{B}/n_\gamma = 2.74 \times 10^{-8}\Omega_\text{B}h^2$ ($\Omega_\text{B} \equiv \rho_\text{B}/\rho_\text{c}$) is in good agreement with the value deduced from observations of CMB anisotropies generated at a redshift of $z \sim 10^3$ when the primordial plasma recombines and requires that baryons (more precisely, nucleons) contribute only $\Omega_\text{B} = 0.012 - 0.025)h^{-2}$, i.e. most of the matter in the universe must be non-baryonic. Moreover this concordance is an extremely powerful constraint on new physics, e.g. it requires that, barring conspiracies, the strengths of all the fundamental interactions (which *together* determine the n/p ratio at decoupling) cannot have been significantly different (more than a few per cent) from their values today. Furthermore,

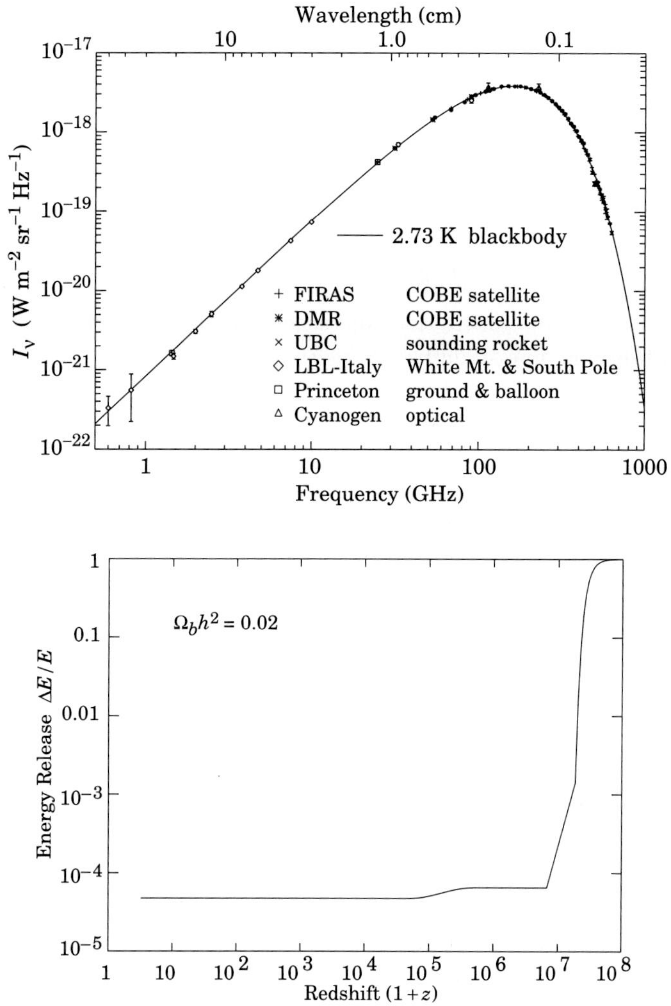

Fig. 1. The spectrum of the cosmic microwave background, demonstrating the excellent fit to a blackbody; as shown in the bottom panel, this imposes severe constraints on any deposition of entropy in the universe back to the thermalization epoch [8].

the dominant energy density in the universe at that epoch must have been radiation — photons and three species of (light) neutrinos. This rules out for example the interesting possibility that there has *always* been a cosmological constant Λ of $\mathcal{O}(H^2)$, since according to the Friedmann equation (4), this is equivalent (taking $k = 0$) to a significant renormalization of the

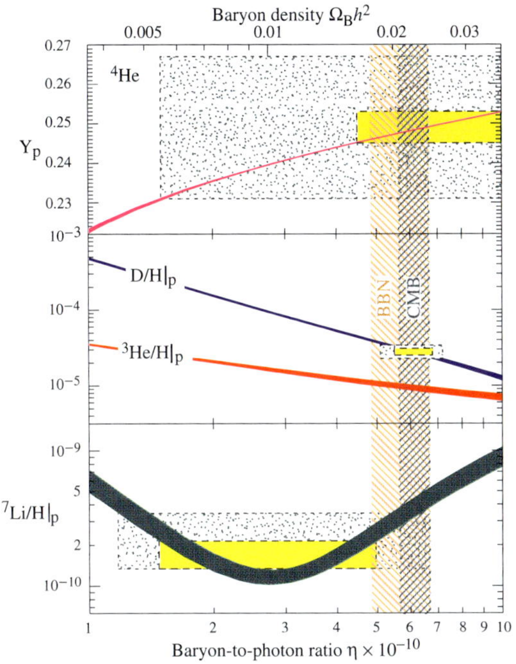

Fig. 2. The abundances of ^{4}He, D, ^{3}He and ^{7}Li as predicted by the standard model of Big Bang nucleosynthesis as a function of $\eta_{10} \equiv \eta/10^{-10}$ [9]. Boxes indicate the observed light element abundances (smaller boxes: 2σ statistical errors; larger boxes: $\pm 2\sigma$ statistical and systematic errors); the narrow vertical band indicates the CMB measure of the cosmic baryon density.

Planck scale (i.e. Newton's constant) which would be in conflict with the observed light element abundances.

These are two of the 'pillars' that the standard Big Bang cosmology is based on and they provide a secure understanding of the thermal history back to when the universe was hot enough to melt nuclei at an age of ~ 1 s. We turn now to a detailed discussion of the third 'pillar', viz. the Hubble expansion, which has been probed back only to a redshift of $\mathcal{O}(1)$ but this of course encompasses most of the actual time elapsed since the Big Bang.

2.1. *The age of the universe and the Hubble constant*

Advances in astronomical techniques have enabled radioactive dating to be performed using stellar spectra. Figure 3 shows the detection of the

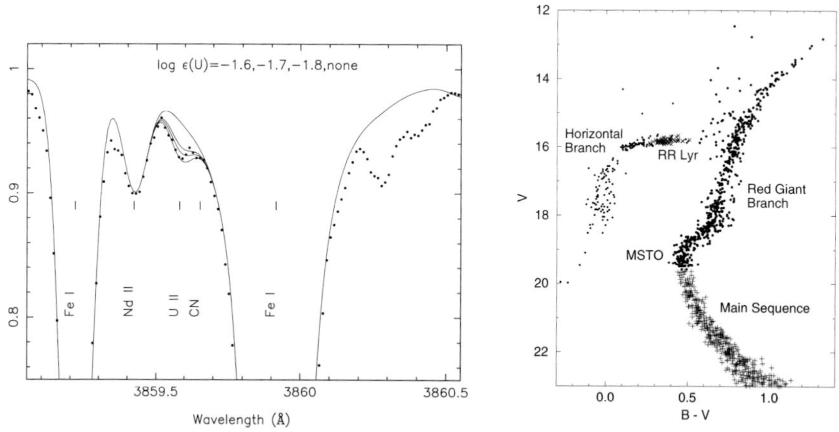

Fig. 3. Detection of ^{238}U in the old halo star CS31802-001; synthetic spectra for three assumed values of the abundance are compared with the data [13]. The right panel shows the color (B-V) versus the magnitude (V) for stars in a typical globular cluster M15; the age is deduced from the luminosity of the (marked) 'main sequence turn-off' point [14].

386 nm line of singly ionized ^{238}U in an extremely metal-poor (i.e. very old) star in the halo of our Galaxy [13]. The derived abundance, log(U/H) $= -13.7 \pm 0.14 \pm 0.12$ corresponds to an age of 12.5 ± 3 Gyr, consistent with the age of 11.5 ± 1.3 Gyr for the (oldest stars in) globular clusters inferred, using stellar evolution models, from the observed Hertzprung–Russell diagram [14]. To this must be added ~ 1 Gyr, the estimated epoch of galaxy/star formation, to obtain the age of the universe.

For the Big Bang cosmology to be valid this age must be consistent with the expansion age of the universe derived from measurement of the present Hubble expansion rate. The *Hubble Space Telescope* Key Project [15] has made direct measurements of the distances to 18 nearby spiral galaxies (using Cepheid variables) and used these to calibrate five secondary methods which probe further; all data are consistent with $H_0 = 72 \pm 3 \pm 7$ km s^{-1} Mpc^{-1}, as shown in Fig. 4. It has been argued however that the Key Project data need to be corrected for local peculiar motions using a more sophisticated flow model than was actually used, and also for metallicity effects on the Cepheid calibration — this would lower the value of H_0 to 63 ± 6 km s^{-1} Mpc^{-1} [16]. Even smaller values of H_0 are also obtained by 'physical' methods such as measurements of time delays in gravitationally lensed systems, which bypasses the traditional 'distance ladder' and probes to far deeper distances than the Key Project. At present ten multiply-imaged quasars have well measured time delays; taking the lenses

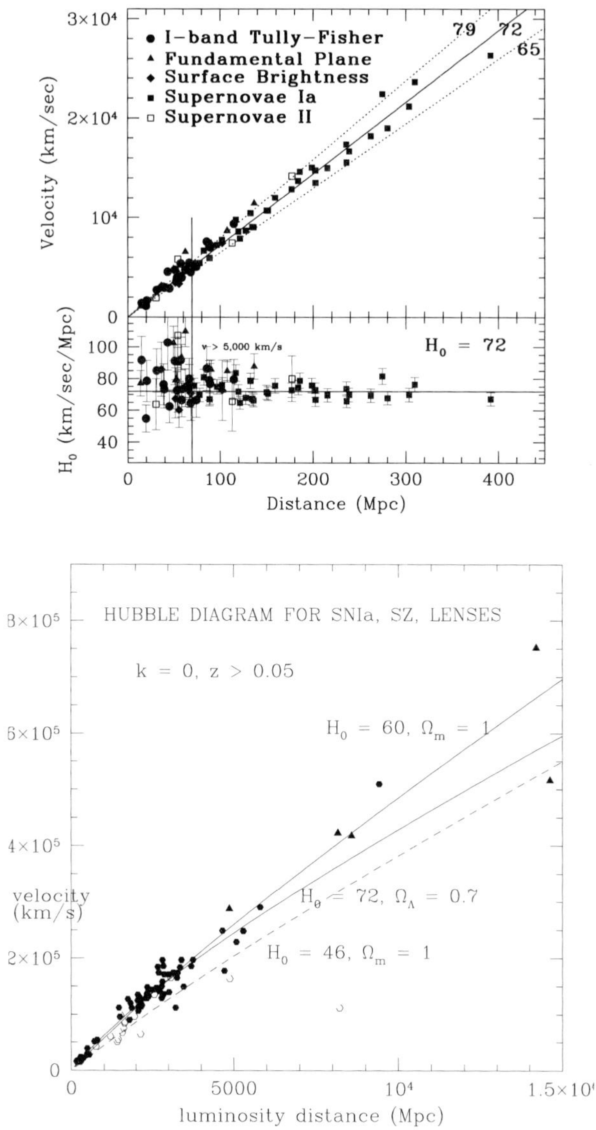

Fig. 4. Hubble diagram for Cepheid-calibrated secondary distance indicators; the bottom panel shows how fluctuations due to peculiar velocities die out with increasing distance [15]. The right panel shows deeper measurements using SNe Ia (filled circles), gravitational lenses (triangles) and the Sunyaev–Zeldovich effect (circles), along with some model predictions [19].

to be isothermal dark matter halos yields $H_0 = 48 \pm 3$ km s^{-1} Mpc^{-1} [17]. Measurements of the Sunyaev–Zeldovich effect in 41 X-ray emitting galaxy clusters also indicate a low value of $H_0 \sim 61 \pm 3 \pm 18$ km s^{-1} Mpc^{-1} for a $\Omega_{\rm m} = 0.3$, $\Omega_\Lambda = 0.7$ universe, dropping further to $H_0 \sim 54$ km s^{-1} Mpc^{-1} if $\Omega_{\rm m} = 1$ [18]. Both these models imply an acceptable age for the universe, taking these uncertainties into account.

2.2. *The deceleration parameter*

The most exciting observational developments in recent years have undoubtedly been in measurements of the deceleration parameter $q \equiv {\rm d}H^{-1}/{\rm d}t - 1 = \frac{\Omega_{\rm m}}{2} - \Omega_\Lambda$. This has been found to be *negative* through deep studies of the Hubble diagram of Type Ia supernovae (SNe Ia) pioneered by the *Supernova Cosmology Project* [20] and the *High-z SN Search Team* [21]. Their basic observation was that distant supernovae at $z \sim 0.5$ are $\Delta m \sim 0.25$ mag (corresponding to $10^{\Delta m/2.5} - 1 \simeq 25\%$) fainter than would be expected for a decelerating universe such as the $\Omega_{\rm m} = 1$ Einstein–deSitter model. This has been interpreted as implying that the expansion rate has been *speeding up* since then, thus the observed SNe Ia are actually further away than expected. Note that the measured apparent magnitude m of a source of known absolute magnitude M yields the 'luminosity distance':

$$m - M = 5 \log \left(\frac{d_{\rm L}}{\rm Mpc} \right) + 25, \quad d_{\rm L} = (1+z) \int_0^z \frac{{\rm d}z'}{H(z')}, \tag{8}$$

which is sensitive to the expansion history, hence the cosmological parameters. According to the second Friedmann equation (4) an accelerating expansion rate requires the dominant component of the universe to have *negative* pressure. The more mundane alternative possibility that the SNe Ia appear fainter because of absorption by intervening dust can be constrained since this would also lead to characteristic reddening, unless the dust has unusual properties [22]. It is more difficult to rule out that the dimming is due to evolution, i.e. that the distant SNe Ia (which exploded over 5 GYr ago!) are intrinsically fainter by $\sim 25\%$ [23]. Many careful analyses have been made of these possibilities and critical reviews of the data have been given [24].

Briefly, SNe Ia are observationally known to be a rather homogeneous class of objects, with intrinsic peak luminosity variations $\lesssim 20\%$, hence particularly well suited for cosmological tests which require a 'standard

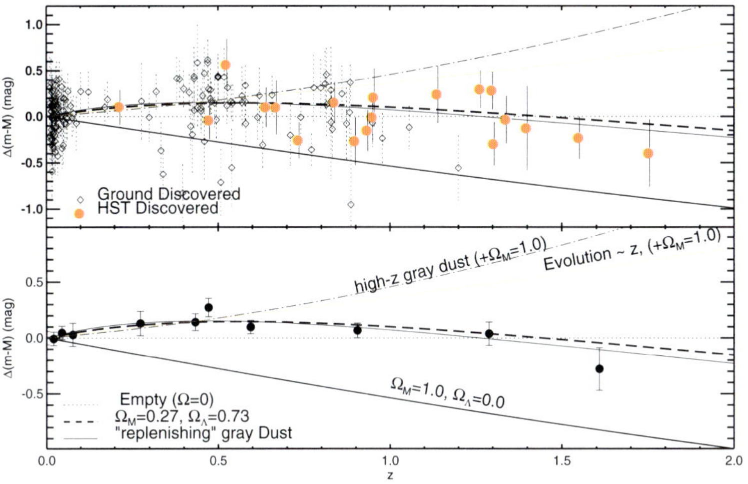

Fig. 5. The residual Hubble diagram of SNe Ia relative to the expectation for an empty universe, compared to models; the bottom panel shows weighted averages in redshift bins [28].

candle' [25]. They are characterized by the absence of hydrogen in their spectra [26] and are believed to result from the thermonuclear explosion of a white dwarf, although there is as yet no "standard model" for the progenitor(s) [27]. However, it is known (using nearby objects with independently known distances) that the time evolution of SNe Ia is tightly correlated with their peak luminosities such that the intrinsically brighter ones fade faster — this can be used to make corrections to reduce the scatter in the Hubble diagram using various empirical methods such as a 'stretch factor' to normalize the observed apparent peak magnitudes [20] or the 'Multi-color Light Curve Shape' method [21]. Such corrections are *essential* to reduce the scatter in the data sufficiently so as to allow meaningful deductions to be made about the cosmological model.

Figure 5 shows the magnitude-redshift diagram of SNe Ia obtained recently by the *Supernova Search Team* [28] — this uses a carefully compiled 'gold set' of 142 SNe Ia from ground-based surveys, together with 14 SNe Ia in the range $z \sim 1 - 1.75$ discovered with the *HST*. The latter are *brighter* than would be expected if extinction by dust or simple luminosity evolution is responsible for the observed dimming of the SNe Ia upto $z \sim 0.5$, and thus support the earlier indication of an accelerating cosmological expansion. However, alternative explanations such as luminosity evolution proportional to lookback time, or extinction by dust which is maintained at a constant density are still possible. Moreover for reasons to do with how SNe Ia are detected, the dataset consists of approximately equal subsamples with redshifts above and below $z \sim 0.3$. It has been noted that this is also the redshift at which the acceleration is inferred to begin and that if these subsets are analyzed *separately*, then the 142 ground-observed SNe Ia are consistent with deceleration; only when the 14 high-z SNe Ia observed by the *HST* are included is there a clear indication of acceleration [29]. Clearly further observations are necessary particularly at the poorly sampled intermediate redshifts $z \sim 0.1 - 0.5$, as is being done by the *Supernova Legacy Survey* [30] and *ESSENCE* [31]; there is also a proposed space mission — the *Supernova Acceleration Probe* [32].

3. The Spatial Curvature and the Matter Density

Although the first indications for an accelerating universe from SNe Ia were rather tentative, the notion that dark energy dominates the universe became widely accepted rather quickly [33]. This was because of two independent lines of evidence which also suggested that there is a substantial cosmological constant. The first was that contemporaneous measurements of degree-scale angular fluctuations in the CMB by the *Boomerang* [34] and *MAXIMA* [35] experiments provided a measurement of the sound horizon (a 'standard ruler') at recombination [36] and thereby indicated that the curvature term $\kappa \simeq 0$, i.e. the universe is spatially flat. The second was that, as had been recognized for some time, several types of observations indicate that the amount of matter which participates in gravitational clustering is significantly less than the critical density, $\Omega_{\rm m} \simeq 0.3$ [37]. The cosmic sum rule then requires that there be some form of 'dark energy', unclustered on the largest spatial scales probed in the measurements of $\Omega_{\rm m}$, with an energy density of $1 - \Omega_{\rm m} \simeq 0.7$. This was indeed consistent with the value of $\Omega_\Lambda \sim 0.7$ suggested by the SNe Ia data [20, 21] leading to the widespread identification of the dark energy with vacuum energy. In fact all data are consistent with $w = -1$ i.e. a cosmological constant, hence the model is termed ΛCDM (since the matter content must mostly be cold dark matter (CDM) given the constraint from BBN on the baryonic component).

Subsequently a major advance has come about with precision measurements of the CMB anisotropy by the *WMAP* satellite, and of the power spectrum of galaxy clustering by the *2dFGRS* and *SDSS* collaborations. The paradigm which these measurements test is that the early universe underwent a period of inflation which generated a gaussian random field of of small density fluctuations ($\delta\rho/\rho \sim 10^{-5}$ with a nearly scale-invariant 'Harrison–Zeldovich' spectrum: $P(k) \propto k^n, n \simeq 1$), and that these grew by gravitational instability in the sea of (dark) matter to create the large-scale structure (LSS) as well as leaving a characteristic anisotropy imprint on the CMB. The latter are generated through the oscillations induced when the close coupling between the baryon and photon fluids through Thomson scattering is suddenly reduced to zero as the universe turns neutral at $z \sim 1000$ [36]. The amplitudes and positions of the resulting 'acoustic peaks' in the angular power spectrum of the CMB are sensitive to the cosmological parameters and it was recognized that precision measurements of CMB anisotropy can thus be used to determine these accurately [38]. However, in practice, there are many 'degeneracies' in this exercise because of the 'prior' assumptions that have to be made [39]. An useful analogy is to see the generation of CMB anisotropy and the formation of LSS as a sort of cosmic scattering experiment, in which the primordial density perturbation is the "beam", the universe itself is the "detector" and its matter content is the "target" [40]. In contrast to the situation in the laboratory, neither the properties of the beam, nor the parameters of the target or even of the detector are known — only the actual "interaction" may be taken to be gravity. In practice, therefore assumptions have to be made about the nature of the dark matter (e.g. 'cold' non-relativistic or 'hot' relativistic?) and about the nature of the primordial perturbation (e.g. adiabatic or isocurvature?) as well as its spectrum, together with further 'priors' (e.g. the curvature parameter k or the Hubble constant h) before the cosmological density parameters can be inferred from the data.

Nevertheless as Fig. 6 shows, the angular spectrum of the all-sky map of the CMB by *WMAP* is in impressive agreement with the expectation for a flat ΛCDM model, assuming a power-law spectrum for the primordial (adiabatic only) perturbation [41]. The fitted parameters are $\Omega_{\rm B}h^2 = 0.024 \pm 0.001$, $\Omega_{\rm m}h^2 = 0.14 \pm 0.02$, $h = 0.72 \pm 0.05$, with $n = 0.99 \pm 0.04$ so it appears that this does herald the dawn of "precision cosmology". Even more impressive is that the prediction for the matter power spectrum (obtained by convoluting the primordial perturbation with the CDM 'transfer function') is in good agreement with the 2dFGRS measurement of the power spectrum of galaxy clustering [42] if there is no 'bias' between the clustering of galaxies and of CDM. Subsequent studies using the power spectrum from *SDSS* [43] and also from spectral observations of the Lyman-α 'forest' (intergalactic gas clouds) [44] have

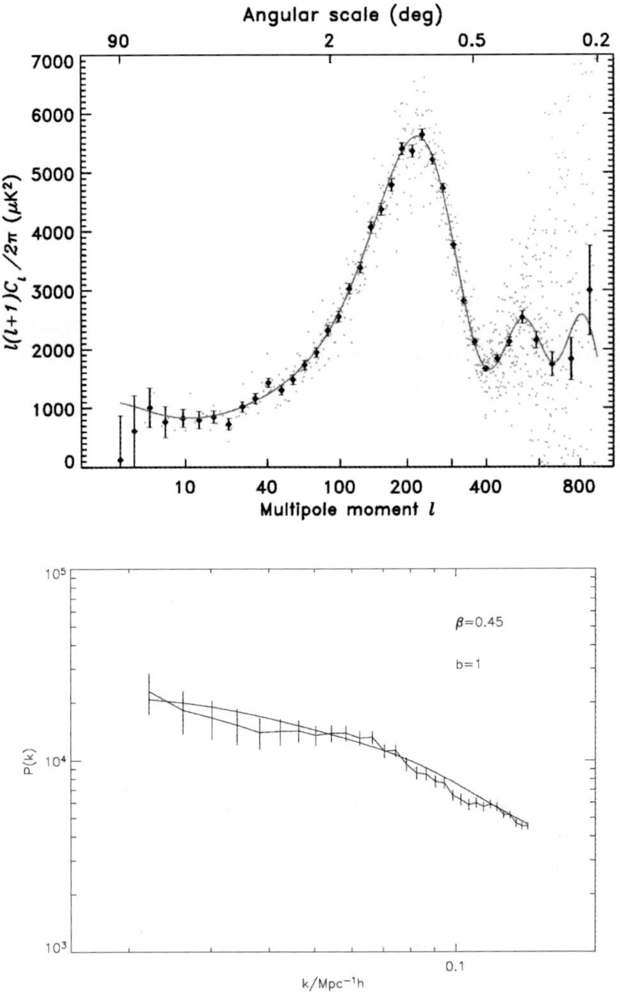

Fig. 6. Angular power spectrum of the CMB measured by WMAP (gray dots are the unbinned data) and the fit to the ΛCDM model; the bottom panel shows the predicted matter power spectrum compared with the *2dFGRS* data assuming no bias between galaxies and dark matter [41].

confirmed these conclusions and improved on the precision of the extracted parameters. Having established the consistency of the ΛCDM model, such analyses also provide tight constraints e.g. on a 'hot dark matter' (HDM) component which translates into a bound on the summed neutrino masses of $\sum m_\nu < 0.42$ eV.

It must be pointed out however that cosmological models *without* dark energy can fit exactly the same data by making different assumptions for the 'priors'. For example, an Einstein–deSitter model is still allowed if the Hubble parameter is as low as $h \simeq 0.46$ and the primordial spectrum is not scale-free but undergoes a change in slope from $n \simeq 1$ to $n \simeq 0.9$ at a wavenumber $k \simeq 0.01$ Mpc^{-1} [19]. To satisfactorily fit the LSS power spectrum also *requires* that the matter not be pure CDM but have a HDM component of neutrinos with (approximately degenerate) mass 0.8 eV (i.e. $\sum m_\nu = 2.4$ eV) which contribute $\Omega_\nu \simeq 0.12$. An alternative to a sharp break in the spectrum is a "bump" in the range $k \sim 0.01 - 0.1$ Mpc^{-1}, such as is expected in models of 'multiple inflation' based on supergravity [45]. Although such models might appear contrived, it must be kept in mind that they do fit all the precision data (except the SNe Ia Hubble diagram) without dark energy and that the degree to which parameters must be adjusted pales into insignifance in comparison with the fine-tuning required of the cosmological constant in the ΛCDM model!

4. Conclusions

Thus for the moment we have a 'cosmic concordance' model with $\Omega_{\mathrm{m}} \sim 0.3, \Omega_\Lambda \sim 0.7$ which is consistent with all astronomical data but has *no* explanation in terms of fundamental physics. One might hope to eventually find explanations for the dark matter (and baryonic) content of the universe in the context of physics beyond the Standard Model but there appears to be little prospect of doing so for the apparently dominant component of the universe — the dark energy which behaves as a cosmological constant. Cosmology has in the past been a data-starved science so it has been appropriate to consider only the simplest possible cosmological models in the framework of general relativity. However, now that we are faced with this serious confrontation between fundamental physics and cosmology, it is surely appropriate to reconsider the basic assumptions (homogeneity, ideal fluids, trivial topology, ...) or even possible alternatives to general relativity.

General relativity has of course been extensively tested, albeit on relatively small scales. Nevertheless the standard cosmology based on it gives a successful account of observations back to the BBN era [46]. However, it is possible that the ferment of current theoretical ideas, especially concerning the possibility that gravity may propagate in more dimensions than matter, might suggest modifications to gravity which are significant in the cosmological context [47, 48]. Astronomers are of course entitled to, and will continue to, analyze their data in terms of well-established physics and treat the cosmological constant as just one among the parameters specifying a cosmological model. However, it is important for it to be recognized that "Occam's razor" does not really apply to the construction of such models, given that there is no physical understanding of the key ingredient Λ.

Landu famously said *"Cosmologists are often wrong, but never in doubt"*. The situation today is perhaps better captured by Pauli's enigmatic remark — the present interpretation of the data may be *"...not even wrong"*. However, we are certainly not without doubt! The crisis posed by the recent astronomical observations is not one that confronts cosmology alone; it is the spectre that haunts any attempt to unite two of the most successful creations of 20th century physics — quantum field theory and general relativity. It is quite likely that the cosmological constant which Einstein allegedly called his *"biggest blunder"* will turn out to be the catalyst for triggering a new revolution in physics in this century.

References

[1] Einstein, A. [1917] Sitzungsber. Preuss. Akad. Wiss. phys.-math. Klasse VI, **142**.

[2] For a historical perspective, see, Straumann, N. [2003] in *Séminaire Poincaré: Vacuum Energy — Renormalization*, eds. Duplantier, B. and Rivasseu, V. (Birkhäuser-Verlag) p. 7.

[3] Weinberg, S. [1989] *Rev. Mod. Phys.* **61**, 1.

[4] See, e.g., Zwirner, F., in *International Europhysics Conference on High Energy Physics, Brussels* (World-Scientific, 1996) p. 943.

[5] For reviews, see e.g. Sahni, V. and Starobinsky, A. A. [2000] *Int. J. Mod. Phys. D* **9**, 373; Peebles, P. J. E. and Ratra, B. [2003] *Rev. Mod. Phys.* **75**, 559; Padmanabhan, T. [2003] *Phys. Rept.* **380**, 235.

[6] For discussions, see, e.g. Witten, E., arXiv:hep-ph/0002297, arXiv:hep-th/0106109; Weinberg, S., arXiv:astro-ph/0005265; Dine, M., arXiv:hep-th/0107259; Nobbenhuis, S., arXiv:gr-qc/0411093, Trivedi, S. P. [2004] *Pramana* **63**, 777.

[7] Mather, J. C., Fixsen, D. J., Shafer, R. A., Mosier, C. and Wilkinson, D. T. [1999] *Astrophys. J.* **512**, 511.

[8] See the review on Cosmic microwave background by Smoot, G. F. and Scott, D. [2002] in *Review of Particle Physics*, Hagiwara, K. *et al.*, *Phys. Rev. D* **66**, 010001.

[9] See the review on Big bang nucleosynthesis by Fields, B. D. and Sarkar, S. [2006] in *Review of Particle Physics*, Yao, W.-M. *et al.*, *J. Phys. G* **33**, 1.

[10] See e.g. Ge, J. A., Bechtold, J. and Black, J. H. [1997] *Astrophys. J.* **474**, 67; Srianand, R., Petitjean, P. and Ledoux, C. [2000] *Nature* **408**, 931; Molaro, P., Levshakov, S. A., Dessauges-Zavadsky, M. and D'Odorico, S. [2002] *Astron. Astrophys.* **381**, L64.
[11] Hoyle, F., Burbidge, G. and Narlikar, J. V. [1993] *Astrophys. J.* **410**, 437; [1995] *Proc. Roy. Soc. Lond. A* **448**, 191.
[12] Peebles, P. J. E., Schramm, D. N., Turner, E. L. and Kron, R. G. [1991] *Nature* **352**, 769.
[13] Cayrel, R. *et al.* [2001] *Nature* **409**, 691.
[14] Chaboyer, B. [1998] *Phys. Rept.* **307**, 23.
[15] Freedman, W. L. *et al.* [2001] *Astrophys. J.* **553**, 47.
[16] Rowan-Robinson, M. [2001] in *Third Int. Conf. Identification of Dark Matter*, eds. Spooner, N. and Kudryavtsev, V. (World Scientific); [2002] *Mon. Not. Roy. Astron. Soc.* **332**, 352.
[17] Kochanek, C. S. and Schechter, P. L. [2004] in *Measuring and Modeling the Universe*, ed. Freedman, W. (Cambridge University Press) p. 117.
[18] Reese, E. D. [2004] in *Measuring and Modeling the Universe*, ed. Freedman, W. (Cambridge University Press) p. 138.
[19] Blanchard, A., Douspis, M., Rowan-Robinson, M. and Sarkar, S. [2003] *Astron. Astrophys.* **412**, 35.
[20] Perlmutter, S. *et al.* [1999] *Astrophys. J.* **517**, 565.
[21] Riess, A. G. *et al.* [1998] *Astron. J.* **116**, 1009.
[22] Aguirre, A. N. [1999] *Astrophys. J.* **525**, 583.
[23] Drell, P. S., Loredo, T. J. and Wasserman, I. [2000] *Astrophys. J.* **530**, 593.
[24] Leibundgut, B. [2000] *Astron. Astrophys. Rev.* **10**, 179; [2001] *Ann. Rev. Astron. Astrophys.* **39**, 67; Filippenko, A. V. [2004] in *Measuring and Modeling the Universe*, ed. Freedman, W. (Cambridge University Press) p. 270.
[25] Branch, D. [1998] *Ann. Rev. Astron. Astrophys.* **36**, 17.
[26] Fillipenko, A. V. [1997] *Ann. Rev. Astron. Astrophys.* **35**, 309.
[27] Hillebrandt, W. and Niemeyer, J. C. [2000] *Ann. Rev. Astron. Astrophys.* **38**, 191.
[28] Riess, A. G. *et al.* [2004] *Astrophys. J.* **607**, 665.
[29] Choudhury, T. R. and Padmanabhan, T. [2005] *Astron. Astrophys.* **429**, 807.
[30] Astier, P. *et al.* [2006] *Astron. Astrophys.* **447**, 31.
[31] Sollerman, J. *et al.*, arXiv:astro-ph/0510026.
[32] Albert, J. *et al.*, arXiv:astro-ph/0507459.
[33] Bahcall, N., Ostriker, J. P., Perlmutter, S. and Steinhardt, P. J. [1999] *Science* **284**, 1481.
[34] P. de Bernardis *et al.* [2000] *Nature* **404**, 955.
[35] Hanany, S. *et al.* [2000] *Astrophys. J.* **545**, L5.
[36] For reviews, see e.g. Hu, W., Sugiyama, N. and Silk, J. [1997] *Nature* **386**, 37; Subramanian, K. [2005] *Curr. Sci.* **88**, 1068.
[37] Peebles, P. J. E., astro-ph/0102327.
[38] Bond, J. R., Crittenden, R., Davis, R. L., Efstathiou, G. and Steinhardt, P. J. [1994] *Phys. Rev. Lett.* **72**, 13; Jungman, G., Kamionkowski, M., Kosowsky, A. and Spergel, D. N. [1996] *Phys. Rev. D* **54**, 1332.

[39] Bond, J. R., Efstathiou, G. and Tegmark, M. [1997] *Mon. Not. Roy. Astron. Soc.* **291**, L33; Efstathiou, G. and Bond, J. R. [1999] *Mon. Not. Roy. Astron. Soc.* **304**, 75.
[40] Sarkar, S. [2005] *Nucl. Phys. Proc. Suppl.* **148**, 1.
[41] Spergel, D. N. *et al.* [2003] *Astrophys. J. Suppl.* **148**, 175.
[42] Percival, W. J. *et al.* [2001] *Mon. Not. Roy. Astron. Soc.* **327**, 1297.
[43] Tegmark, M. *et al.* [2004] *Phys. Rev. D* **69**, 103501.
[44] Seljak, U. *et al.* [2005] *Phys. Rev. D* **71**, 103515.
[45] Adams, J. A., Ross, G. G. and Sarkar, S. [1997] *Nucl. Phys. B* **503**, 405; Hunt, P. and Sarkar, S. [2004] *Phys. Rev. D* **70**, 103518; Hunt, P., Morgan, A. and Sarkar, S., to appear.
[46] Peebles, P. J. E. [2005] in *Proc. 17th Int. Conf. General Relativity and Gravitation*, eds. Florides, P. *et al.* (World Scientific), p. 106.
[47] Dvali, G. R., Gabadadze, G. and Porrati, M. [2000] *Phys. Lett. B* **485**, 208.
[48] Carroll, S. M. *et al.* [2005] *Phys. Rev. D* **71**, 063513.

CHAPTER 12

Gravitational Radiation — In Celebration of Einstein's *Annus Mirabilis*

****

B. S. SATHYAPRAKASH

School of Physics and Astronomy, Cardiff University, Cardiff, CF24 3AA, UK

Two of Einstein's 1905 papers were on special theory of relativity. Although general relativity was to come a decade later, it was special relativity that was responsible for the existence of wave-like phenomena in gravitation. A hundred years after the discovery of special relativity we are poised to detect gravitational waves and the detection might as well come from another inevitable and exotic prediction of relativity, namely black holes. With interferometric gravitational wave detectors taking data at unprecedented sensitivity levels and bandwidth, we are entering a new century in which our view of the Universe might be revolutionized yet again with the opening of the gravitational window. The current generation of interferometric and resonant mass detectors are only the beginning of a new era during which the gravitational window could be observed by deploying pulars and microwave background radiation.

1. Introduction

We are celebrating 100 years of Einstein's *Annus Mirabilis* 1905, during which he published four papers on three subjects, thereby laying the foundation for *quantum theory,* the *theory of Brownian motion,* and the *special theory of relativity.* Each of these subjects has revolutionized our world view but the story is not over yet. Though it was not until 1915 that he founded the general theory of relativity, it was special relativity that is responsible for the existence of gravitational radiation. The waves have eluded direct detection so far but there is little doubt today about their existence thanks to spectacular observations of the decay in the orbital period of Galactic binary neutron stars [1, 2].

Gravity is generated by the densities and currents of the energy and momentum of matter; if they change then the gravitational field changes. In Einstein's theory of gravity, unlike in Newtonian gravity, changes in the distribution of a source cannot travel instantaneously but at the speed of light. Indeed, the field equations of general relativity admit wave-like solutions that are in many ways similar to electromagnetic (EM) radiation but with important differences. Gravitational interaction is Universal and, although weak, gravity is nonlinear. Universality of gravitation means that one cannot infer the influence of gravitational waves by watching an isolated particle in space, as opposed to EM radiation whose influence on a single charged particle can be inferred. One would need at least two particles, just as one would in Einstein's gedanken lift-experiment to infer the presence of the Earth's gravitational field. The weakness of the interaction means that on the one hand it will be very difficult to generate gravitational waves in the laboratory; only catastrophic astronomical events involving massive accelerations of bulk matter, as opposed to EM waves which are produced by accelerated charged particles, can produce significant amplitudes of the radiation. On the other hand, the radiation carries the true signature of the emitting source, be it the core of a neutron star or a supernova, the quasi-normal mode oscillations of a black hole, or the birth of the Universe, thereby making it possible to observe phenomena and objects that are not directly accessible to the electromagnetic, neutrino or the cosmic-ray window. This is unlike EM waves which interacts very strongly with matter and therefore imprint on the radiation from a source is the characteristics of its 'surface' rather than the core. Nonlinearity of gravitational waves implies that the waves interact with the source resulting in a rich structure in the shape of the emitted signals that will be useful for testing relativity in new ways.

Gravitational radiation can be characterized by a dimensionless amplitude which is a measure of the deformation in space caused by the wave as it passes by. For instance, if two masses are initially separated by a distance ℓ, a wave of amplitude h causes a change in length $\delta\ell = h\ell/2$. Typical astronomical events, say a binary black hole merger at 100 Mpc, would have an amplitude $h \sim 10^{-23}$ at a frequency $\sim$ 100 Hz. Detectors that are currently in operation will be able to observe such events which are expected to occur about once per year.

In the rest of this article we will discuss astronomical sources of gravitational waves. This article chiefly deals with compact objects, namely neutron stars (NS) and black holes (BH). Unless specified otherwise we shall assume that a NS has a mass of $M = 1.4M_\odot$ and radius $R = 10$ km, and by a stellar mass BH we shall mean a black hole of mass $M = 10M_\odot$. We shall assume a flat Universe with a cold dark matter density of $\Omega_M = 0.3$, dark energy of $\Omega_\Lambda = 0.7$, and a Hubble constant of $H_0 = 65\,\mathrm{km\,s^{-1}\,Mpc^{-1}}$. We shall use a system of units in which $c = G = 1$, which means $1M_\odot \simeq 5 \times 10^{-6}\,\mathrm{s} \simeq 1.5\,\mathrm{km}$, $1\,\mathrm{Mpc} \simeq 10^{14}\,\mathrm{s}$.

I will begin this article with a brief overview of gravitational wave (GW) theory and their interaction with matter and how that is used in the construction of detectors. The main focus of the article will be the astronomical sources of gravitational waves.

2. Gravitational Wave Theory — A Brief Overview

A key point of Newton's gravity is that because the potential satisfies Poisson equation, $\nabla^2\varphi(t,\,\mathbf{x}) = 4\pi\rho(t,\,\mathbf{x})$, there are no retardation effects. Since there are no time-derivatives involved on the LHS and the *same* time t appears both on the LHS and the RHS, any change in the distribution of density at the location of the source would *instantaneously* change the potential at the remote field point. In Einstein's theory the metric components of the background spacetime are the gravitational potentials and they satisfy a "wave" equation and hence there will be retardation effects. This is explicitly seen in the linearized version of Einstein's equations.

Under the assumption of weak gravitational fields one can assume that the background metric $g_{\alpha\beta}$ of spacetime to be only slightly different from the Minkowski metric $\eta_{\alpha\beta} = \mathrm{Diag}(-1,\,1,\,1,\,1) : g_{\alpha\beta} = \eta_{\alpha\beta} + h_{\alpha\beta}$. $|h_{\alpha\beta}| \ll 1$, called the *metric perturbation*, describes the departure of the spacetime from flatness. Under the assumption of weak gravitational fields, general relativistic equations for the metric perturbation can be linearized to obtain a set of wave equations for the metric perturbation: $\Box\overline{h}_{\alpha\beta} = 16\pi T_{\alpha\beta}$, where $2\overline{h}_{\alpha\beta} \equiv 2h_{\alpha\beta} - \eta_{\alpha\beta}h^\mu_\mu$, and $\Box$ is the wave operator, $\Box \equiv \eta^{\alpha\beta}\partial_\alpha\partial_\beta$, and $T_{\alpha\beta}$ is the energy-momentum tensor. These equations have the formal solution

$$\overline{h}_{\alpha\beta}(t,\mathbf{x}) = 4\int \frac{T_{\alpha\beta}(t - |\mathbf{x} - \mathbf{x}'|,\,\mathbf{x}')\,\mathrm{d}^3x'}{|\mathbf{x} - \mathbf{x}'|}. \tag{1}$$

Now the metric perturbation at the field point $\mathbf{x}$ at time t is determined by the configuration of the source $T_{\alpha\beta}$ at a retarded time $t - |\mathbf{x} - \mathbf{x}'|$ (recall that we are using units in which the speed of light is unity). Hence disturbances in the source travel only at a finite speed. Indeed, any non-stationary source $T_{\alpha\beta}$ will give rise to wave-like solutions for the potentials $\overline{h}_{\alpha\beta}$, which extract energy, momentum and angular-momentum from the source, propagating at the speed of light.

2.1. *Effect of gravitational waves on matter*

Just as in EM theory, there are only two independent polarizations of the field, denoted as h_+ (h-plus) and $h_\times$ (h-cross), and not ten components as one might expect from the tensorial nature of $h_{\alpha\beta}$. This is because the theory is generally covariant and there are gauge degrees of freedom.

A wave of plus-polarization traveling along, say, z-axis continuously deforms a circular ring of beads in the x-y plane, taking the ring from a circle to an ellipse with its semi-major axis first oriented along, say, the x-axis after one-quarter of the wave, then along the y-axis after three-quarter and so on. Monitoring the distance from the center of the ring to the beads at the ends of two orthogonal radial directions can best measure the deformation of the ring. This is the principle behind a laser interferometer antenna. If the ring's original radius was R the semi-major and semi-minor axes of the ellipse would be $(1+h/2)R$ and $(1-h/2)R$, respectively. A wave of cross-polarization has similar effect but the whole pattern of deformation gets rotated by $\pi/4$. This contrasts with the oscillatory response of a single charged particle to an EM wave and the two polarizations of light are at an angle of $\pi/2$ relative to each other.

Gravitational wave interferometers are quadrupole detectors with a good sky coverage. A single antenna, except when it is a spherical resonant detector, cannot determine the polarization state of a *transient* wave or the direction to the source that emits the radiation. Interferometers or resonant bars do not measure the two polarizations separately but rather a linear combination of the two given by:

$$h(t) = F_+(\theta, \varphi, \psi)h_+(t) + F_\times(\theta, \varphi, \psi)h_\times(t), \tag{2}$$

where F_+ and $F_\times$ are the antenna patterns. To infer the direction (θ, φ) to the source, the polarization amplitudes $(h_+, h_\times)$, and the polarization angle ψ, it is necessary to make five measurements, which is possible with three interferometers: each interferometer gives a response, say $h_1(t)$, $h_2(t)$ and $h_3(t)$, and one can infer two independent delays, say $t_1 - t_2$, and $t_2 - t_3$, in the arrival times of the transient at the antennas. Therefore, a network of antennas, geographically widely separated so as to maximize the time delays and hence improve directionality, is needed for GW observations. Moreover, detecting the same event in two or more instruments helps to remove the non-Gaussian and non-stationary backgrounds, while adding a greater degree of confidence to the detection of an event. In the case of continuous waves and stochastic backgrounds the motion of the detector relative to the source causes a Doppler modulation of the response which can be de-convolved from the data to fully reconstruct the wave (as one would in the case of a point source) or reconstruct a map of the sky (as one would in the case of a stochastic background).

2.2. *Amplitude, luminosity and frequency*

The amplitude h and luminosity $\mathcal{L}$ of a source of GW is given in terms of the famous quadrupole formula (see for details, e.g. [3]):

$$h_{mn}(t, \mathbf{r}) = \frac{2}{r}\ddot{\mathcal{I}}_{mn}(t - \mathbf{r}), \qquad \mathcal{L} = \frac{1}{5}\langle \dddot{\mathcal{I}}_{mn} \dddot{\mathcal{I}}^{mn} \rangle, \tag{3}$$

where an overdot denotes derivative with respect to time; angular brackets denote a suitably defined averaging process (say, over a period of the GW); $\mathcal{I}_{mn}$ is the *reduced* (or trace-free) quadrupole moment tensor which is related to the usual quadrupole tensor $I^{mn} \equiv \int T^{00} x^m x^n \mathrm{d}^3 x$, via $\mathcal{I}_{mn} \equiv I_{mn} - \delta_{mn} I^k_{\ k}/3$. In simple terms, for a source of size R, mass M and angular frequency ω, located at a distance r from Earth,

$$h \sim \epsilon_h \frac{M}{r} R^2 \omega^2, \qquad \mathcal{L} \sim \epsilon_{\mathcal{L}} M^2 R^4 \omega^6 . \tag{4}$$

where $\epsilon_{h,\mathcal{L}}$ are dimensionless efficiency factors that depend on the orientation of the system relative to the observer (in the case of h only) and how *deformed* from spherical symmetry the system is. $\epsilon_{h,\mathcal{L}} \sim 1$ for ideally oriented and highly deformed sources. The amplitude of the waves, just as in the case of electromagnetic radiation, decreases as inverse of the distance to the source. However, there is a crucial difference between EM and GW observations that is worth pointing out: Let r_l be the largest distance from which an EM or a GW detector can observe standard candles. In the case of EM telescopes r_l is limited by the smallest flux observable, which falls off as the inverse-square of the distance. This is because astronomical EM radiation is the superposition of waves emitted by a large number of microscopic sources, each photon with its own phasing; we cannot follow each wave separately but only a superposition of many of them. This, of course, is the reason why in conventional astronomy the number counts of standard candles increase as $r_l^{3/2}$. In the case of GW, signals we expect to observe are emitted by the coherent bulk motion of large masses and hence it is possible to observe each cycle of the wave as it passes through the antenna. Indeed, one can fold many wave cycles together to enhance the visibility of the signal buried in noise, provided the shape of the signal is known before-hand. Because we can follow the amplitude of a wave the number of sources which an antenna can detect increases as r_l^3.

For a self-gravitating system, say a binary system of two stars of masses m_1 and m_2 (total mass $M = m_1 + m_2$ and symmetric mass ratio $\eta = m_1 m_2/M^2$), the linear velocity v and angular velocity ω are related to the size R of the system via Keplar's laws: $\omega^2 = M/R^3$, $v^2 = M/R$. It turns out that the efficiency factors for such a system are $\epsilon_h = 4\eta\, C$, $\epsilon_{\mathcal{L}} = 32\eta^2/5$, so that

$$h \simeq 4\eta\, C \frac{M}{r}\frac{M}{R}, \qquad \mathcal{L} \simeq \frac{32}{5}\eta^2 v^{10}, \qquad f_{\rm GW} = 2 f_{\rm orb}\,, \tag{5}$$

where $C \sim 1$ is a constant that depends on the orientation of the source relative to the detector, $f_{\rm GW}$ is the GW frequency which is equal to twice the orbital frequency $f_{\rm orb}$.[1] The above relations imply that the amplitude of a source is greater the more compact it is and the luminosity is higher from a source that is more relativistic. The factor to covert the luminosity from $G = c = 1$ units to conventional units is $\mathcal{L}_0 \equiv c^5/G \simeq= 3.6 \times 10^{59}$ erg s^{-1}. Since $v < 1$, $\mathcal{L}_0$ denotes the best luminosity a source could ever have and generally $\mathcal{L} \ll \mathcal{L}_0$.

3. Gravitational Wave Detector Projects

There are chiefly two types of GW detectors that are currently in operation taking sensitive data: (i) *resonant bars* and (ii) *laser interferometers.* The sensitivity of a detector is defined in terms of the smallest discernible dimensionless strain caused by an astronomical source against background noise of the instrument. Because a GW antenna can follow the phase of GW, the sensitivity of an antenna is given in terms of the amplitude noise spectrum as a function of frequency and is measured in Hz$^{-1/2}$. Figure 1, bottom panel, shows in solid lines the design sensitivity goals of three generations of ground-based interferometers (shown here for the American initial and advanced LIGO, and a possible third generation European detector

[1] For a binary consisting of two equal masses the configuration of the system is identical on rotation by π, rather than 2π, radians. This is the reason why the frequency of GW is twice the orbital frequency. In general, the wave would contain the orbital frequency and its harmonics with twice the orbital frequency being the dominant.

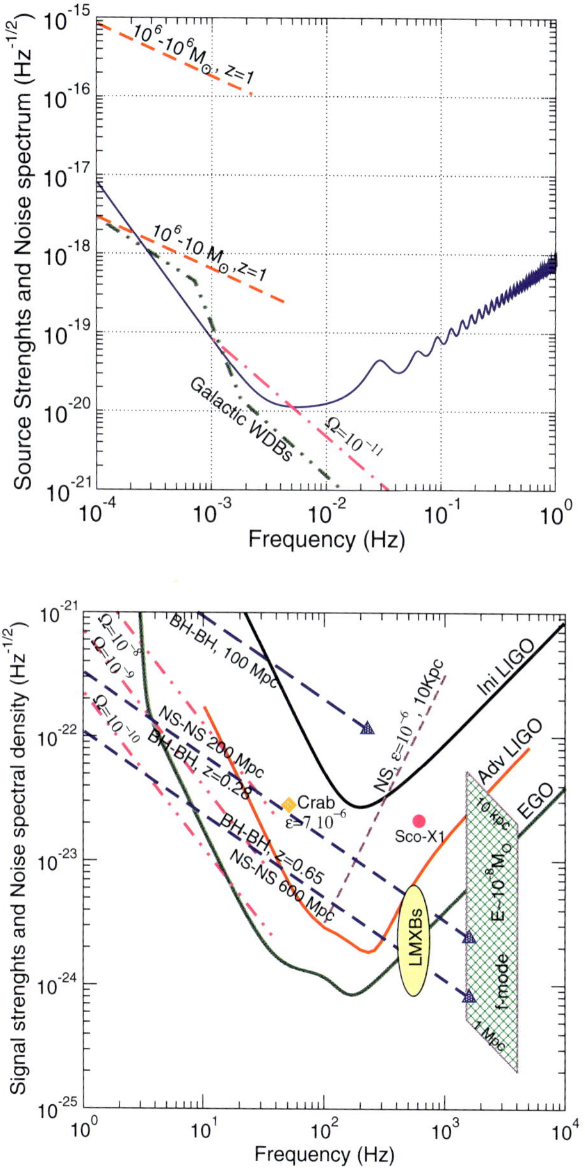

Fig. 1. Amplitude noise spectrum (in $\mathrm{Hz}^{-1/2}$) of space-based LISA (top panel) and three generations of ground-based interferometers, initial LIGO, advanced LIGO, and EGO (bottom panel). Also plotted on the same graph are the source strengths for an archetypal binary, continuous and stochastic radiation in the same units. A source will be visible in a network of three interferometers if it is roughly five times above the noise curve.

EGO). The top panel shows the same for the space-based LISA. Also plotted in Fig. 1 are source strengths to be discussed in Sec. 4

3.1. *Ground-based interferometers*

Interferometers operate in a broad band with a peak sensitivity at frequency of ~ 150 Hz (see Schutz [4] for a fuller description). In a laser interferometric antenna the tidal deformation caused in the two arms of a Michelson interferometer is sensed as a shift in the fringe pattern at the output port of the interferometer. The sensitivity of such a detector is limited at low frequencies (10–40 Hz) by anthropogenic sources and seismic disturbances, at intermediate frequencies (40–300 Hz) by thermal noise of optical and suspended components, and at high frequencies (> 300 Hz) by photon shot noise. Three key technologies have made it possible to achieve the current level of sensitivities: (1) An optical layout that makes it possible to recycle the laser light exiting the interferometer and build effective powers that are 1000's of times larger than the input thereby mitigating the photon shot noise. This technique allows us to operate the interferometer either in the wide band mode (as in Fig. 1), or with a higher sensitivity in a narrow band of about 10–50 Hz centered at a desired frequency, say 300 Hz, but at the cost of worsened sensitivity over the rest of the band. This latter mode of operation is called *signal recycling* and is particularly useful for observing long-lived continuous wave sources. (2) Multiple suspension systems that filter the ground motion and keep the mirrors essentially free from seismic disturbances. (3) Monolithic suspensions that help isolate the thermal noise to a narrow frequency band.

There are currently six long baseline detectors in operation: The American Laser Interferometer Gravitational-Wave Observatory (LIGO) [5], which is a network of three detectors, two with 4 km arms and one with 2 km arms, at two sites (Hanford, Washington and Livingstone, Louisiana), the French-Italian VIRGO detector with 3 km arms at Pisa [6], the British-German GEO600 [7] with 600 m arms at Hannover and the Japanese TAMA300 with 300 m arms in Tokyo [8]. Australia has built a 80 m test facility with a plan to build a km-size detector sometime in the future. The American LIGO detectors are already close to their design sensitivity. Figure 2 shows the sensitivity of these detectors during the fourth science run (called S4). The GEO600 and Virgo detectors are expected to operate at their design sensitivity during the course of the next year. LIGO and GEO600 began a year long science run in late 2005, to be joined in due

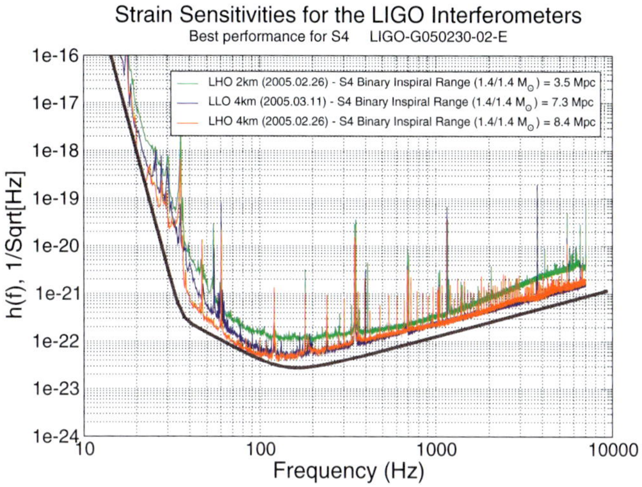

Fig. 2. Amplitude noise spectrum (in $\mathrm{Hz}^{-1/2}$) of LIGO interferometers during the fourth science run S4. At this time the LIGO instruments were roughly within a factor of 2 of their design goal, but are currently running at design sensitivity.

course by TAMA300 and Virgo. The data from the worldwide network will be the best ever, providing a real chance for a first direct detection.

Plans are well underway both in Europe and the USA to build, by 2010–2013, the next generation of interferometers that are 10–15 times more sensitive than the initial interferometers or to enhance the sensitivity in the kHz region where we can expect to observe neutron star cores and quasi-normal modes of stellar mass black holes. With a peak sensitivity of $h \sim 10^{-24}\,\mathrm{Hz}^{-1/2}$ the advanced LIGO and VIRGO detectors will be able to detect NS ellipticities in the range 10^{-6}–10^{-8} from sources in our Galaxy, BH–BH binaries of total mass $50M_\odot$ at a redshift of $z \sim 0.5$, and stochastic background at the level of $\Omega_{\mathrm{GW}} \sim 10^{-9}$. The high-frequency upgrade of GEO600, called GEO-HF, is being designed to observe normal mode oscillations of neutron stars, believed to result from star quakes and the root cause of glitches in pulsar timing when the frequency of the pulsar seems to suddently increase.

In the longer term, over the next 10 to 15 years, we might see the development of third generation GW antennas. The sensitivity of the current and next generation instruments is still far from the fundamental limitations of a ground-based detector: The gravity gradient noise at low frequencies due to natural (winds, clouds, earth quakes) and anthropogenic causes, and the quantum uncertainty principle of mirror position at high frequencies. A detector subject to only these limitations requires the development of new optical and cryogenic techniques that form the foundation of a third generation GW detector that is currently undergoing a design study.

3.2. *Space-based interferometers*

Laser Interferometer Space Antenna (LISA) is a joint ESA-NASA project to develop a space-based gravitational wave detector. The plan is to put in space three spacecraft in heliocentric orbit, 60 degrees behind the Earth, in an equilateral triangular formation of size 5 million km [10], scheduled to be ready by 2015. LISA's sensitivity is limited by difficulties with long time-scale ($> 10^5$ s) stability, photon shot-noise at high frequencies ($\sim 10^{-3}$ Hz) and large size ($> 10^{-1}$ Hz). LISA will be able to observe Galactic, extra-Galactic and cosmological point sources as well as stochastic backgrounds from different astrophysical populations and perhaps from certain primordial processes. Feasibility studies are now underway to assess the science case and technology readiness for covering the frequency gap of LISA and ground-based detectors. The *Deci-Hertz Interferometer Gravitational-Wave Observatory* (DECIGO) [11] by the Japanese team and the *Big-Bang Observer* (BBO), a possible follow-up of LISA [12], are aimed as instruments to observe the primordial GW background and to answer cosmological questions on the expansion rate of the Universe and dark energy.

4. Sources of Gravitational Waves

Gravitational wave detectors will unveil dark secrets of the Universe by detecting sources in extreme physical environs: strong nonlinear gravity, relativistic motion, extremely high density, temperature and magnetic fields, to list a few. We shall focus our attention on *compact objects* (in isolation or in binaries) and *stochastic backgrounds*.

4.1. *Compact binaries*

An astronomical binary consisting of a pair of compact objects, i.e. a neutron star and/or a black hole, is called a compact binary. GWs are emitted as the two bodies in such a system orbit around each other, carrying away, in the process, the orbital angular momentum and energy from the system. The dissipation of energy and angular momentum leads to a slow inspiral of the stars towards each other. The inspiral phase of the evolution, which can be treated adiabatically due to low loss of energy (at least initially), is very well modeled to a high order in post-Newtonian (PN) theory and we have a precise theoretical prediction of the phasing of the emitted radiation given the masses of the two bodies. Thus a compact binary is an astronomer's ideal standard candle [13]: A parameter called the *chirp mass* $\mathcal{M} \equiv \eta^{2/3}M$, completely fixes the absolute luminosity of the system. Hence, by observing GW from a binary we can measure the luminosity distance to the source provided the source *chirps,* that is the orbital frequency changes, by as much as $1/T$ during an observational period T, thereby enabling the measurement of the chirp mass. Consequently, it will be possible to accurately measure cosmological parameters and their variation as a function of red-shift.

The dynamics of a compact binary consists of three phases: (i) *inspiral,* (ii) *plunge* and (iii) *merger.* In the following we will discuss each in turn.

(i) The *early inspiral phase:* This is the phase in which the system spends 100's of millions of years and the power emitted in GW is low. This phase can be treated using linearized Einstein's equations and post-Newtonian theory with the energy radiated into gravitational waves balanced by a loss in the binding energy of the system. The emitted GW signal has a characteristic shape with its amplitude and frequency slowly increasing with time and is called a *chirp* waveform. Formally, the inspiral phase ends at the *last stable orbit* (LSO) when the effective potential of the system undergoes a transition from having a well-defined minimum to the one without a minimum, after which stable orbits can no longer be supported. This happens roughly when the two objects are separated by $R \simeq 6\,GM/c^2$, or when the frequency of GW is $f_{\rm LSO} \simeq 4400\,(M_\odot/M)\,{\rm Hz}$. The signal power drops as $f^{-7/3}$ and the photon shot-noise in an interferometer increases as f^2 beyond about 200 Hz so that it will only be possible to detect a signal in the range from about 10 Hz to 500 Hz (and a narrower bandwidth of 40–300 Hz in initial interferometers) during which the source brightens up half-a-million fold (recall that the luminosity $\propto v^{10} \propto f^{10/3}$).

For $M \lesssim 10M_\odot$, inspiral phase is the only phase sensed by the interferometers and lasts for a duration of $\tau = 5576\,\eta^{-1}\,(M/M_\odot)^{-5/3}\,(f_0/10\,{\rm Hz})^{-8/3}$ s, where f_0 is the frequency at which the observation begins. The phase development of the signal is very well modeled during this epoch and one can employ matched filtering technique to enhance the visibility of the signal by roughly the square root of the number of signal cycles

$N_{\rm cyc} \sim (8/5)\tau f_0$. Since a large number of cycles are available it is possible to discriminate different signals and accurately measure the parameters of the source such as its location (a few degrees each in co-latitude and azimuth) [14], mass (fractional accuracy of 0.05–0.3% in total mass and a factor 10 worse for reduced mass, with greater accuracy for NS than BH), and spin (to within a few percents) [15].

(ii) The *late inspiral, plunge* and *merger phase*: This is the phase when the two stars are orbiting each other at a third of the speed of light and experiencing strong gravitational fields with the gravitational potential being $\varphi = GM/Rc^2 \sim 0.1$. This phase warrants the full nonlinear structure of Einstein's equations as the problem involves strong relativistic gravity, tidal deformation (in the case of BH–BH or BH–NS) and disruption (in the case of BH–NS and NS–NS) and has been the focus of numerical relativists [16] for more than two decades. However, some insights have been gained by the application of advanced mathematical techniques aimed at accelerating the convergence properties of post-Newtonian expansions of the energy and flux required in constructing the phasing of GW [17–19]. This is also the most interesting phase from the point of view of testing nonlinear gravity as we do not yet fully understand the nature of the two-body problem in general relativity. Indeed, even the total amount of energy radiated during this phase is highly uncertain, with estimates in the range 10% [20] to 0.7% [18]. Since the phase is not well-modeled, it is necessary to employ sub-optimal techniques, such as time-frequency analysis, to detect this phase and then use numerical simulations to gain further insights into the nature of the signal.

(iii) The *late merger phase:* This is the phase when the two systems have merged to form either a single NS or a BH, settling down to a quiescent state by radiating the deformations inherited during the merger. The emitted radiation can be computed using perturbation theory and gives the quasi-normal modes (QNM) of BH and NS. The QNM carry a unique signature that depends only on the mass and spin angular momentum in the case of BH, but depends also on the equation-of-state (EOS) of the material in the case of NS. Consequently, it is possible to test conclusively whether or not the newly born object is a BH or NS: From the inspiral phase it is possible to estimate the mass and spin of the object quite precisely and use that to infer the spectrum of normal modes of the BH. The fundamental QNM of GW from a spinning BH, computed numerically and then fitted, is [21] $f_{\rm QNM} = 750[1 - 0.63(1-a)^{0.3}](100M_\odot/M)$ Hz, with a decay time of $\tau = 5.3/[f_{\rm QNM}(1-a)^{0.45}]$ ms, where a is the dimensionless spin parameter of the hole taking values in the range [0, 1]. The signal will be of the form $h(t; \tau, \omega) = h_0 e^{-t/\tau} \cos(2\pi f_{\rm QNM} t)$, $t \geq 0$, h_0 being the amplitude of the signal that depends on how deformed the hole is.

It has been argued that the energy emitted in the form of QNM might be as large as 3% of the system's total mass [20]. By matched filtering, it should be possible to detect QNM resulting from binary mergers of mass in the range 60–$10^3 M_\odot$ at a distance of 200 Mpc in initial LIGO and from $z \sim 2$ in advanced LIGO. In Fig. 1 filled circles (connected by a dotted line) show the amplitude and frequency of QNM radiation from a source at $z = 2$, and total mass 1000, 100 or $10 M_\odot$. Such signals should serve as a probe to test if massive black holes found at galactic cores initially formed as small BHs of $10^3 M_\odot$ and then grew by rapid accretion. Moreover, there is a growing evidence [22] that globular clusters might host BHs of mass $M \sim 10^3 M_\odot$ at their cores. If this is indeed true then the QNM from activities associated with such BHs would be observable in the local Universe, depending on how much energy is released into GW when other objects fall in. EGO could also observe QNM in stellar mass black holes of mass $M \sim$ 10–$20 M_\odot$.

The span of an interferometer to binaries varies with the masses as $\eta^{1/2} M^{5/6}$, greater asymmetry in the masses reduces the span but larger total mass increases the span. In Fig. 3 we have plotted the distance up to which binaries can be seen as a function of the binary's total mass for an equal mass system when including both the inspiral and merger part of the signal. This estimate is based on the effective-one-body approach [18, 23] which predicts 0.7% of the total mass in the merger waves.

4.1.1. *NS–NS binaries*

Double NS coalescences can be seen in initial LIGO to a distance of about 15 Mpc and in advanced LIGO to a distance of 300 Mpc as shown in Fig. 3 (top panel). Based on the observed small population of binary NS that merge within the Hubble time, Kalogera *et al.* [2, 24] conclude that the Galactic coalescence rate is $\sim 1.8 \times 10^{-4}$ yr^{-1}, which would imply an event rate of NS–NS coalescences is 0.25 and 1500 yr^{-1}, in initial and advanced LIGO, respectively. The rates are uncertain by a factor of about 100, however, due largely to the small number of coalescing binaries that are known today. As the spins of NS are very small ($a \ll 1$) and because the two stars would merge well outside the LIGO's sensitivity band, the current state-of-the-art theoretical waveforms [25] will serve as good templates for matched filtering. However, detailed relativistic hydrodynamical simulations (see,

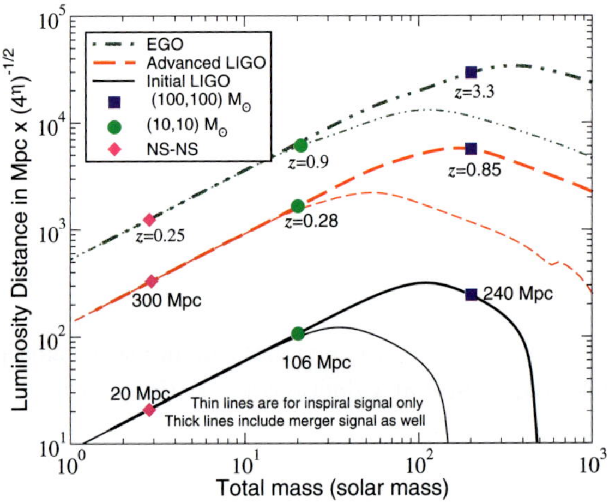

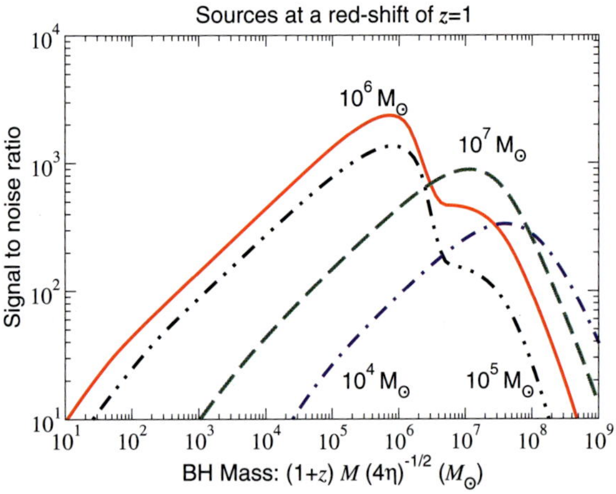

Fig. 3. The top plot shows the span of initial and advanced LIGO and EGO for compact binary sources when including both the inspiral and merger waveforms in our search algorithms. BH mergers can be seen out to a red-shift of $z = 0.28$ in advanced LIGO and $z = 0.9$ in EGO. In the bottom plot we show the SNR achieved by LISA for inspiral signals from supermassive black hole binaries at $z = 1$. The plot shows the SNR for different masses of one of the black holes as a function of the other black hole's mass.

e.g. Ref. [26]) would be needed to interpret the emitted radiation during the coalescence phase, wherein the two stars collide to form a bar-like structure prior to merger. The bar hangs up over a couple of dynamical time-scales to get rid of its deformity by emitting strong bursts of GW. Observing the radiation from this phase should help to deduce the EOS of NS bulk matter. Also, an event rate as large as in advanced LIGO and EGO will be a valuable catalogue to test astronomical predictions, for example if γ-ray bursts are associated with NS–NS and/or NS–BH mergers [27].

4.1.2. *NS–BH binaries*

These are binaries consisting of one NS and one BH and are very interesting from an astrophysical point of view: The initial evolution of such systems can be treated analytically fairly well, however, the presence of a BH with large spin can cause the NS to be whirled around in precessing orbits due to strong spin-orbit coupling. The evolution of such systems is really not very well understood. However, it should be possible to use the "point-mass" approximation in which the NS is treated as a point-particle orbiting a BH, in getting some insight into the dynamics of the system. The evolution will also be complicated by the tidal disruption of the NS before reaching the last stable orbit. It should be possible to accurately measure the onset of the merger phase and deduce the radius of the NS to $\sim 15\%$ and thereby infer its EOS [28].

Advanced interferometers will be sensitive to NS–BH binaries out to a distance of about 650 Mpc. The rate of coalescence of such systems is not known empirically as there have been no astronomical NS–BH binary identifications. However, the population synthesis models give [29] a Galactic coalescence rate in the range 3×10^{-7}–5×10^{-6} yr^{-1}. The event rate of NS–BH binaries will be worse than BH–BH of the same total mass by a factor of $(4\eta)^{3/2}$ since the SNR goes down as $\sqrt{4\eta}$. Taking these factors into account we get an optimistic detection rate of NS–BH of 0.05 and 400 per year in initial and advanced LIGO, respectively.

4.1.3. *BH–BH binaries*

Black hole mergers are the most promising candidate sources for a first direct detection of GW. These sources are the most interesting from the view point of general relativity as they constitute a pair of interacting Kerr spacetimes experiencing the strongest possible gravitational fields before they merge with each other to form a single Kerr BH, and serve as a platform to test general relativity in the nonlinear regime. For instance, one can detect the scattering of GW by the curved geometry of the binary [30], [31], test post-Newtonian theory to a very high order [32] and measure, or place upper limits on, the mass of the graviton to 2.5×10^{-22} eV and 2.5×10^{-26} eV in ground- and space-based detectors, respectively [33]. High SNR events (which could occur once every month in advanced LIGO) can be used to test the full nonlinear gravity by comparing numerical simulations with observations and thereby gain a better understanding of the two-body problem in general relativity. As BH binaries can be seen to cosmological distances, a catalogue of such events compiled by LIGO can be used to measure Cosmological parameters (Hubble constant, expansion of the Universe, dark energy) and test models of Cosmology [27].

The span of interferometers to BH–BH binaries varies from 100 Mpc (with the inspiral signal only) to 150 Mpc (inspiral plus merger signal) in initial LIGO and to a red-shift of $z = 0.28$ in advanced LIGO, and $z = 0.9$ in EGO (cf. Figs. 1 and 3). As in the case of NS–BH binaries, here too there is no empirical estimate of the event rate. Population synthesis models are highly uncertain about the Galactic rate of BH–BH coalescences and predict [29] a galactic rate of 3×10^{-8}–10^{-5} yr^{-1}, which is smaller than the predicted rate of NS–NS coalescences. However, owing to their greater masses, BH–BH event rate in our detectors is larger than NS–NS by a factor $M^{5/2}$ for $M \lesssim 100 M_\odot$. The predicted event rate is a maximum of 1 yr^{-1} in initial LIGO and 500 yr^{-1} to 20 day^{-1} in advanced LIGO.

4.1.4. *Massive black hole binaries*

It is now believed that the center of every galaxy hosts a BH whose mass is in the range 10^6–$10^9 M_\odot$ [34]. These are termed as *massive black holes* (MBH). There is now observational evidence that when galaxies collide the MBH at their nuclei might get close enough to be driven by gravitational radiation reaction and merge within the Hubble time [35]. For a binary with $M = 10^6 M_\odot$ the frequency of GW at the last stable orbit is $f_{\rm LSO} = 4$ mHz, followed by merger from 4 mHz to the QNM at 24 mHz (if the spin of the black holes is negligible). This is in the frequency range of LISA which has been designed to observe the MBH: their formation, merger and activity.

The SNR for MBH–MBH mergers in LISA is shown in Fig. 3. These mergers will appear as the most spectacular events requiring no templates for signal identification, although good models would be needed to extract source parameters. Supermassive black hole mergers can be seen almost wherever they occur; one can observe galaxy mergers throughtout the Universe and address astrophysical questions about the origin, growth and population of MBH. The recent discovery of a MBH binary [35] and the association of X-shaped radio lobes with the merger of MBH [36] has raised the optimism concerning MBH mergers and the predicted rate for MBH mergers is the same as the rate at which galaxies merge, about 1 yr^{-1} out to a red-shift of $z = 5$ [37].

4.1.5. *Smirches*

The MBH environment of our own galaxy is known to constitute a large number of compact objects and white dwarfs. Three-body interaction will occasionally drive these compact objects, white dwarfs and other stars into a capture orbit of the central MBH. The compact object will be captured in an highly eccentric trajectory ($e > 0.99$) with the periastron close to the last stable orbit of the MBH. Due to relativistic frame dragging, for each passage of the apastron the compact object will experience several turns around the MBH in a near circular orbit. Therefore, long periods of low-frequency, small-amplitude radiation will be followed by several cycles of high-frequency, large-amplitude radiation. The apastron slowly shrinks, while the periastron remains more or less at the same location, until the final plunge of the compact object before merger. There will be a lot of structure in the waveforms which arise as a result of a number of different physical effects: Contribution from higher order multipoles, precession of the orbital plane that changes the polarization of the waves observed by LISA, the rich spectrum that results from a precessing elliptic orbit, etc. This complicated structure smears the power in the signal in the time-frequency plane [38] as compared to a sharp chirp from a non-spinning BH binary and for this reason this *spin modulated chirp* is called a *smirch* [39].

As the compact object tumbles down the MBH it will sample the spacetime geometry in which it is moving and the structure of that geometry will be imprint in the GW emitted in the process. By observing smirches, LISA offers a unique opportunity to directly map the spacetime geometry around the central object and test whether or not this structure is in accordance with the expectations of general relativity [40]. Indeed, according to Einstein's theory the geometry of a rotating black hole is uniquely determined to be the Kerr metric involving just two parameters, the mass of the MBH and its spin. Thus, the various multipole-moments of the source are uniquely fixed once we have measured the mass and spin of the BH. With the observed smirch one can basically test whether general relativity correctly describes the spacetime region around a generic BH and if the central object is indeed a BH or some other exotic matter.

The SNR from smirches will be between 10–50 depending on the mass of the central object (cf. Fig. 3) but it might be very difficult to detect them by matched filtering due to their complicated shapes, although the event rate is expected to be rather high. Indeed, a background population of these smirches will cause confusion noise and only sources in the foreground will be visible in LISA. The foreground event rate is somewhat uncertain, ranging from 1–10 yr^{-1} within 1 Gpc [41].

4.2. *Neutron stars*

Neutron stars are the most compact stars in the Universe. With a density of 2×10^{14} g cm^{-3}, and surface gravity $\varphi \equiv M/R \sim 0.1$, they are among the most exotic objects whose composition, equation-of-state and structure, are still largely unknown. Being highly compact they are potential sources of GW. The waves could be generated either from the various normal modes of the star, or because the star has a tiny deformation from spherical symmetry and is rotating about a non-axisymmetric axis, or because there are density inhomogeneities caused by an environment, or else due to certain relativistic instabilities. We will consider these in turn.

4.2.1. *Supernovae and birth of NS*

The birth of a NS is preceded by the gravitational collapse of a highly evolved star or the core collapse of an accreting white dwarf. Type II supernovae (SN) are believed to result in a compact remnant. In any case, if the collapse is non-spherical then GW could carry away some of the binding energy and angular momentum depending on the geometry of the collapse. It is estimated that in a typical SN, GW might extract about 10^{-7} of the total energy [42]. The waves could come off in a burst whose frequency might lie in the range $\sim$ 200–1000 Hz. Advanced LIGO will be able to see such events up to the Virgo supercluster with an event rate of about 30 per year.

4.2.2. *Equation of state and normal modes of NS*

In order to determine the equation of state (EOS) of a neutron star, and hence its internal structure, it is necessary to independently determine its mass and radius. Astronomical observations cannot measure the radius of neutron stars although radio and X-ray observations do place a bound on their mass. Therefore, it is has not been possible to infer the EOS.

Neutron stars will have their own distinct normal modes and GW observations of these modes should resolve the matter here since by measuring the frequency and damping times of the modes it would be possible to infer both the radius and mass of NS. The technique is not unlike helioseismology where observation of normal modes of the Sun has facilitated insights into its internal structure. In other words, GW observations of the normal modes of the NS will allow *gravitational asteroseismology* [43].

Irrespective of the nature of the collapse a number of normal modes will be excited in a newly formed NS. The star will dissipate the energy in these modes in the form of GWs as a superposition of the various normal modes and soon the star settles down to a quiescence state. Normal modes could also be excited in old NS because of the release of energy from star quakes. The strongest of these modes, the ones that are important for GW observations, are the *p*- and *w*-modes for which the restoring forces are the fluid pressure and space-time curvature, respectively. Both of these modes will emit transient radiation which has a generic form of a damped sinusoid: $h(t;\, \nu, \tau) = h_0 \exp(-t/\tau)\cos(2\pi\nu t)$, where h_0 is the amplitude of the wave that depends on the external perturbation that excites the mode and ν and τ are the frequency and damping time of the mode, respectively, and are determined by the mass and radius of the NS for a given EOS and depend on the type of mode excited.

To make an order-of-magnitude estimate let us assume that the mass of the NS is $M_\star = 1.4 M_\odot$ and that its radius is $R_\star = 10\,\mathrm{km}$. For the *p*-modes, which are basically fluid modes, the frequency of the fundamental mode, also called the *f*-mode, is simply the dynamical frequency of the fluid, namely $\nu_f \sim \sqrt{\rho}$, where ρ is the density of the fluid. For a NS of radius $R_\star$ and mass $M_\star$ this corresponds to a frequency of $\sqrt{3M_\star/(4\pi R_\star^3)} \sim 3\,\mathrm{kHz}$. If the star radiates the energy E deposited in the mode at a luminosity $\mathcal{L}$, the damping time of the mode would be $\tau \sim E/\mathcal{L}$. Since $E \propto M_\star^2/R_\star$ and $\mathcal{L} \propto M_\star^2 R_\star^4 \omega^6 = M_\star^5/R_\star^5$, we get $\tau \sim R_\star^4/M_\star^3$. Indeed, detailed mode calculations for various EOS have been fitted to yield the following relations for *f*-modes [43]

$$\begin{aligned} \nu_f &= \left[0.78 + 1.635\left(\frac{M_\star}{R_\star^3}\right)^{1/2}\right] \mathrm{kHz}\,, \\ \tau_f^{-1} &= \frac{M_\star^3}{R_\star^4}\left[22.85 - 14.65\frac{M_\star}{R_\star}\right] \mathrm{s}\,, \end{aligned} \tag{6}$$

and similarly for *w*-modes. The *f*- and *w*-mode frequencies nicely separate into two distinct groups even when considering more than a dozen different

EOS: The f-modes are in the frequency range 1–4 kHz, w-modes are in the range 8–14 kHz, and therefore, detecting a signal at these frequencies places it in one or the other category. The frequency and damping time, together with the relations above, can then be used to fix the radius and mass of the star. Observing several systems should then yield a mass-radius curve which is distinct for each EOS and thereby helps to address the question of NS structure.

The amplitude of f- and w-modes corresponding to 12 different EOS from NS at 10 kpc to 15 Mpc is shown in Fig. 1 (bottom panel) as two shaded regions. In a typical gravitational collapse the amount of energy expected to be deposited in f- or w-modes, $\sim 10^{-8} M_\odot$, makes it impossible to detect them in initial LIGO and barely in advanced LIGO instruments, even for a Galactic source. However, EGO should be able to detect these systems with a high SNR. The event rates for these systems would be at least as large as the supernova rate, i.e. about 0.1–0.01 yr^{-1} in our galaxy, increasing to 10–100 yr^{-1} within the Virgo supercluster.

The same plot can also be used to infer the sensitivity of the detectors to normal modes that might be excited in magnetars or radio pulsars, by noting that the amplitude scales inversely with the distance to the source and directly as the square-root of the energy in the modes. Thus, EGO will be sensitive to normal modes excited with $E \sim 10^{-12} M_\odot$ in the Vela pulsar (300 pc away), which is seen to glitch rather frequently.

4.2.3. *Relativistic instabilities in NS*

NS suffer *dynamical* and *secular* instabilities caused by *hydrodynamical* and *dissipative* forces, respectively. What is of interest to us is the secular instability driven by gravitational radiation. GW emission from a normal mode in a non-spinning NS would always lead to the decay of the mode. However, the situation might reverse under certain conditions: Imagine a NS spinning so fast that a normal mode whose angular momentum (AM) in the star's *rest frame* is *opposite* to its spin, appears to an *inertial observer* to be *co-rotating* with the spin. In the inertial frame, GW extracts positive AM from the mode; therefore the mode's own AM should become more negative. In other words, the amplitude of the mode should grow as a result of GW emission, and hence the instability. The energy for the growth of the mode comes from the rotational energy of the star, which acts like a pump field. Consequently, the star would spin down, the mode's angular momentum with respect to the inertial observer will vanish and thereby halt the instability. It was expected that this instability, called the *CFS instability* [44, 45], might drive the f-modes in a NS unstable, but the star should spin at more than 2 kHz (the smallest f-mode frequency) for this to happen. Moreover, it has been shown that due to viscous damping in the NS fluid the instability would not grow sufficiently large, or sustain for long, to be observable (see e.g. Ref. 43).

It was recently realized [46] that modes originating in current-multipoles, as opposed to mass-multipoles which lead to the f-mode, could be unstable even in a non-spinning NS. These modes, called the r-modes, have received a lot of interest because they could potentially explain why spin frequencies of NS in low-mass X-ray binaries are all clustered in a narrow range of 300–600 Hz or why no NS with spin periods smaller than 1.24 ms have been found. The role of r-modes in these circumstances is as yet inconclusive because the problem involves very complicated astrophysical processes (magnetic fields, differential rotation, superfluidity and superconductivity), microphysics (the precise composition of NS — hyperons, quarks) and full nonlinear physics of general relativity. It is strongly expected that r-modes will be emitted by newly formed NS during the first few months of their birth [43, 47]. The frequency of these modes will be 4/3 of the spin frequency of the star and might be particularly important if the central object in a low-mass X-ray binary is a strange star [48]. The radiation might last for about 300 years and the signal would be detectable in initial LIGO with a few weeks of integration.

4.2.4. *NS environment*

A NS with an accretion disc would be spun up due to transfer of AM from the disc. Further, accretion builds up density inhomogeneities on the NS that could lead to the emission of GW. The resulting radiation reaction torque would balance the accretion torque and halt the NS from spinning up. It has been argued [49] that GW emission could be the cause for spin frequencies of NS in low-mass X-ray binaries to be locked up in a narrow frequency range of 300–600 Hz. It is also possible that r-modes are responsible for the locking up of frequencies instead, in which case the waves would come off at a different frequency [48]. These predictions can be tested with advanced LIGO or EGO as Sco-X1, a nearby low-mass X-ray binary, would produce quite a high SNR (marked as $\triangle$ in Fig. 1, bottom panel).

4.2.5. *Spinning NS with asymmetries*

Our galaxy hosts a population of $\sim 10^8$ NS and they normally spin at high rates (several to 500 Hz) which would induce considerable equatorial bulge and flattening of the poles. The presence of a magnetic field may cause the star to spin about an axis that is different from the symmetry axis leading to a time-varying quadrupole moment [50]. Gravitational waves emitted by such a NS, a distance of $r = 10\,\mathrm{kpc}$ from the Earth, will have an amplitude [51] $h \sim 8 \times 10^{-26} f_{\mathrm{kHz}}^2 \epsilon_{-6}$, where f_{kHz} is the frequency of GW in kHz and ϵ_{-6} is the ellipticity of the star in units of 10^{-6}. Figure 1, bottom panel, plots the signal strength expected from a NS with $\epsilon = 10^{-6}$ at 10 kpc integrated over four months.

The ellipticity of neutron stars is not known but one can obtain an upper limit on it by attributing the observed spin-down rate of pulsars as entirely due to gravitational radiation back reaction, namely that the change in the rotational energy is equal to GW luminosity. The ellipticity of the Crab pulsar inferred in this way is $\epsilon \leq 7 \times 10^{-4}$. The GW amplitude corresponding to this ellipticity is $h \leq 10^{-24}$. Noting that Crab has a spin frequency of 25 Hz (GW frequency of 50 Hz), on integrating the signal for 10^7 s one obtains $h = 3.3 \times 10^{-21}\,\mathrm{Hz}^{-1/2}$, which is easily reachable by LIGO. It is unlikely that the ellipticity is so large and hence the GW amplitude is probably much less. However, seeing Crab at a hundredth of this ellipticity is quite good with advanced LIGO as indicated by a diamond in Fig. 1 (Note that Crab is at 2 kpc, so with an ellipticity of $\epsilon = 7 \times 10^{-6}$ the signal strength would be 35 times higher than the NS line.)

4.3. *Stochastic background*

Incoherent superposition of radiation from a population of background sources and/or quantum processes in the early Universe [29] might produce stochastic signals that would fill the whole space. By detecting such a stochastic signal we can gain knowledge about the underlying populations and physical processes. A network of antennas can be used to discover stochastic signals buried under the instrumental noise background. It is expected that the instrumental background will not be common between two geographically well-separated antennas. Thus, by cross-correlating the data from two detectors we can eliminate the background and filter the interesting stochastic signal. However, when detectors are not co-located the SNR builds only over wavelengths longer than twice the distance between antennas, which in the case of the two LIGO antennas means over frequencies $\lesssim 40$ Hz [52]. The visibility of a stochastic signal integrated over a period T and bandwidth f only increases as $(fT)^{1/4}$ since cross-correlation uses a 'noisy' filter. But the noise in a bandwidth f is $\sqrt{fS_h(f)}$. Thus, the signal effectively builds up as $(T/f)^{1/4}$.

4.3.1. *Astronomical backgrounds*

There are thousands of white dwarf binaries in our galaxy with their period in the range from a few hours to ~ 100 seconds, which is in the frequency band of the space-based LISA. Each binary will emit radiation at a single frequency, but over an observation period T each frequency bin of width $\Delta f = 1/T$ will be populated by many sources. Thus, unless the source is nearby it will not be possible to detect it amongst the *confusion* background created by the underlying population. However, a small fraction of this background population might be detectable as strong foreground sources. The parameters of many white dwarfs are known so well that we can precisely predict their SNRs in LISA and thereby use them to calibrate the antenna. In Fig. 1 the curve labeled WDB (top panel) is the expected confusion noise from Galactic white dwarfs [29, 53]. NS and BH populations do not produce a large enough background to be observable. Note that the white dwarf background imposes a limitation on the sources we can observe in the frequency region from 0.3 mHz to about 1 mHz — the region where we expect smirches to occur.

4.3.2. *Primordial background*

A cosmological background should have been created in the very early Universe and later amplified, as a result of parametric amplification, by its coupling to the background gravitational field [29]. Imprint on such a background are the physical conditions that existed in the early Universe, as also the nature of the physical processes that produced the background. Observing such a background is, therefore, of fundamental importance as this is the only way we can ever hope to directly witness the birth of the Universe. The cosmic microwave background, which is our firm proof of the hot early phase of the Universe, was strongly coupled to baryons for 350,000 years after the big bang and therefore the signature of the early Universe is erased from it. The GW background, on the other hand, is expected to de-couple from the rest of matter 10^{-24} s after the big bang, and would, therefore, carry uncorrupted information about the conditions in the very early Universe.

The strength of stochastic GW background is measured in terms of the fraction $\Omega_{\rm GW}$ of the energy density in GW as compared to the critical density needed to close the Universe and the amplitude of GW is given by [51]: $h = 8 \times 10^{-19}\,\Omega_{\rm GW}^{1/2}/f$, for $H_0 = 65\ {\rm km\,s^{-1}\,Mpc}$. By integrating for 10^7 s, over a bandwidth f, we can measure a background density at $\Omega_{\rm GW} \simeq 4 \times 10^{-5}$ in initial LIGO, 5×10^{-9} in advanced LIGO, 10^{-10} in EGO and 10^{-10} in LISA (cf. Fig. 1 dot-dashed curves marked $\Omega_{\rm GW}$). In the standard inflationary model of the early Universe, the energy density expected in GW is $\Omega_{\rm GW} \lesssim 10^{-15}$, and this will not be detected by future ground-based detectors or LISA. However, space missions currently being studied (DECIGO/BBO) to exploit the astrophysically quiet band of 10^{-2}–1 Hz might detect the primordial GW and unveil the origin of the Universe.

5. Conclusions

Direct detection of gravitational radiation will be an extremely important step in opening a new window on the Universe. Interferometric and resonant mass detectors will play a key role in this step. Gravitational radiation they are expected to observe should facilitate both quantitatively and qualitatively new tests of Einstein's gravity including the measurement of the speed of gravitational waves, and hence (an upper limit on) the mass of the graviton, polarization states of the radiation, nonlinear effects of general relativity untested in solar system or Hulse–Taylor binary pulsar observations, uniqueness of axisymmetric spacetimes, the mystery of galaxy formation, and so on.

The future holds other ways of observing cosmic gravitational waves. Radio astronomy promises to make it possible to observe both primordial gravitational waves and point sources. Polarization spectra of the cosmic background radiation will carry the signature of the primordial waves. Future radio antennas (most notably, the Square Kilometer Array) will monitor tens of thousands of pulsars which can be used as an array of high-precision clocks to observe the gravitational Universe.

References

[1] Weisberg, J. M. and Taylor, J. H. [2003] The relativistic binary pulsar B1913+16, in *Radio Pulsars*, eds. Bailes, M., Nice, D. J. and Thorsett, S. E. (ASP. Conf. Series).

[2] Kalogera, V. *et al.* [2003] The cosmic coalescence rates for double neutron star binaries, astro-ph/0312101.

[3] Schutz, B. F. [1985] *A First Course in General Relativity* (Cambridge University Press, Cambridge).

[4] Schutz, B. F. [1999] *Class. Quant. Grav.* **16**, A131.

[5] Abramovici, A. *et al.* [1992] *Science* **256**, 325.

[6] Caron, B. *et al.* [1997] *Class. Quant. Grav.* **14**, 1461.

[7] Lück, H. *et al.* [1997] *Class. Quant. Grav.* **14**, 1471.

[8] Tsubono, K. [1995] in *First Edoardo Amaldi Conf. Gravitational Wave Experiments* (World Scientific, Singapore), p. 112.

[9] EURO — Europe's Third Generation Gravitational Wave Observatory, `http://www.astro.cf.ac.uk/geo/euro/`

[10] Bender, P. *et al.* [1998] *LISA: Pre-Phase A Report*, MPQ 208 (Max-Planck-Institut für Quantenoptik, Garching, Germany, Second Edition, July 1998).

[11] Takahashi, R. and Nakamura, T. [2003] *ApJ* **596**, L231–L234.

[12] Big Bang Observer, 2003, `http://universe.gsfc.nasa.gov/program/bbo.html`

[13] Schutz, B. F. [1986] *Nature* **323**, 310.

[14] Jaranowski, P., Kokkotas, K. D., Krolak, A. and Tsegas G. [1996] *Class. Quant. Grav.* **13**, 1279.

[15] Cutler, C. and Flanagan, É. É. [1994] *PRD* **49**, 2658.

[16] Bruegmann, B. [2000] *Annalen Phys.* **9**, 227–246.

[17] Buonanno, A. and Damour, T. [1999] *PRD* **59**, 084006.

[18] Buonanno, A. and Damour, T. [2000] *PRD* **62**, 064015.

[19] Damour, T., Iyer, B. R. and Sathyaprakash, B. S. [1998] *PRD* **57**, 885.

[20] Flanagan, É. É. and Hughes, S. [1998] *PRD* **57**, 4535.

[21] Echeverria, F. [1989] *PRD* **40**, 3194.

[22] Gerssen, J. *et al.* [2002] *ApJ* (in press) astro-ph/0210158.

[23] Damour, T., Iyer, B. R. and Sathyaprakash, B. S. [2001] *PRD* **63**, 044023.

[24] Burgay, M. *et al.* [2004] *Nature* **426**, 531–533.

[25] Blanchet, L. [2002] *Living Rev. Rel.* **5**, 3.

[26] Kawamura, M., Oohara, K. and Nakamura, T. [2003] *General Relativistic Numerical Simulation on Coalescing Binary Neutron Stars and Gauge-Invariant Gravitational Wave Extraction,* astro-ph/0306481.

[27] Finn, L. S. [1996] *PRD* **53**, 2878.

[28] Vallisneri, M. [2000] *PRL* **84**, 3519.

[29] Grishchuk, L. P. *et al.* [2001] *Phys. Usp.* **44**, 1 (2001).

[30] Blanchet, L. and Sathyaprakash, B. S. [1994] *Class. Quant. Grav.* **11**, 2807.

[31] Blanchet, L. and Sathyaprakash, B. S. [1995] *PRL* **74**, 1067.

[32] Arun, K. G., Iyer, B. R., Qusailah, M. S. S., Sathyaprakash, B. S. [2006] *PRD* **74**, 024006; *ibid.*, [2006] *Class. Quant. Grav.* **23**, L37–L43.

[33] Will, C. M. [1998] *PRD* **57**, 2061.

[34] Rees, M. J. [1997] *Class. Quant. Grav.* **14**, 1411.

[35] Komossa, S. *et al.* [2003] *ApJ* **582**, L15–L20.

[36] Merritt, D. and Ekers, R. D. [2002] *Science* **297**, 1310–1313.

[37] Haehnelt, M. G. [1998] in *Laser Interferometer Space Antenna*, ed. Folkner, W. M., **456**, 45 (AIP Conference Proceedings, Woodbury, NY).
[38] Sathyaprakash, B. S. [2003] Problem of searching for spinning black hole binaries, *Proc. XXXVIII Recontres de Moriond*, March 24–29 (in press).
[39] Sathyaprakash, B. S. and Schutz, B. F. [2003] *Class. Quant. Grav.* **20**, S209.
[40] Ryan, F. D. [1995] *PRD* **52**, 5707.
[41] Phinney, S. [2001] private communication.
[42] Müller, E. [1997] in *Relativistic Gravitation and Gravitational Radiation*, eds. Lasota, J.-P. and Marck J.-A. (Cambridge University Press, Cambridge).
[43] Andersson, N. and Kokkotas, K. [2001] *IJMPD* **10**, 381.
[44] Chandrasekhar, S. [1970] *PRL* **24**, 611.
[45] Friedman, J. L. and Schutz, B. F. [1978] *ApJ* **222**, 281.
[46] Andersson, N. [1998] *ApJ* **502**, 708.
[47] Lindblom, L., Owen, B. J. and Morsink, S. M. [1998] *PRL* **80**, 4843.
[48] Andersson, N., Jones, D. I. and Kokkotas, K. [2001] *MNRAS* **337**, 1224.
[49] Bildsten, L. [1998] *ApJ Letters* **501**, 89.
[50] Cutler, C. [2002] *PRD* **66**, 084025.
[51] Thorne, K. S. [1987] Gravitational radiation, in *300 Years of Gravitation*, eds. Hawking, S. W. and Isreal, W. (Cambridge University Press, Cambridge), p. 330.
[52] Allen, B. and Romano, J. D. [1999] *PRD* **59**, 102001.
[53] Hils, D., Bender, P. L. and Webbink, R. F. [1990] *ApJ* **360**, 75.
[54] Cutler, C. and Thorne, K. S. [2001] An overview of gravitational wave sources, in *Proc. 16th Int. Conf. General Relativity and Gravitation* (*GR*16), Durban, South Africa, 15–21 July 2001; e-Print Archive: gr-qc/0204090.

CHAPTER 13

Albert Einstein: Radical Pacifist and Democrat

****

T. JAYARAMAN

Institute of Mathematical Sciences, Chennai 600 113, India

We draw attention here to the radical political grounding of Einstein's pacifism. We also drescribe some less commonly known aspects of his commitment to civil liberties, particularly in the context of the anti-left hysteria and anti-racism current in the United States of the late 1940s and 1950s. We also examine briefly his views on socialism.

To Einstein himself, his scientific work was always to be at the core of his being, the very definition of his persona. Nowhere is this clearer than in the substance and style of his Autobiographical Notes that he wrote for the volume *Albert Einstein: Philosopher-Scientist*, edited by P. A. Schlipp [1]. The note, which Einstein begins by describing it as an 'my own obituary', has no reference even to the bare facts of his life, apart from brief comments on his education and the intellectual influences of his childhood and youth. It is entirely devoted to a short account of his main work and the philosophical and scientific questions that led up to them. He interrupts a critique of Newtonian physics in the note to remark: ' "Is this supposed to be an obituary?" the astonished reader will likely ask. I would like to reply: essentially yes. For the essential in the being of a man of my type lies precisely in what he thinks and how he thinks, not in what he does or suffers.'

In this account there is not even the briefest mention of his views on any subject other than the scientific questions that occupied him throughout his scientific career. But this stance, that virtually dismisses his social and political views, belies Einstein's considerable and not inconsequential engagement with many of the major social and political issues of his day. Einstein's broad involvement in public affairs was undoubtedly part of the reason for the iconic status that he was to attain. While it was Einstein's science that propelled him to international fame, Einstein remained in the public eye not in the least due to his regular and willing intervention in public affairs.

Einstein's life spanned some of the most tumultuous years of a turbulent century. His *annus mirabilis* was a decade before the First World War. By the time he died, two World Wars had run their tragic course, the first atomic bombs had been tested, and global politics was dominated by the Cold War that split the world into two camps, armed to their teeth with nuclear weapons, that confronted each other across the globe. The years between the two World Wars was very much the era of the socialist revolution even if most of them were short-lived attempts with the sole exception of the Russian revolution that gave rise to the Soviet Union. Fascism rose to power and was defeated in the course of the Second World War, but not before perpetrating the Holocaust and extracting a grim toll of human lives, particularly in Eastern Europe. The socialist wave continued in the immediate aftermath of the Second World War, even if its appeal was far more dominant in the Third World. A wave of national independence movements in the first half of the century ended the old style of colonial rule that had held sway over a significant section of the world's population even though colonial powers did not cede their powers before sowing the seeds of conflicts in areas such as the Middle East that continue to take their toll even today.

In his reactions to these developments and the political and social issues that they raised, Einstein did not articulate at length an unified political and social philosophy, unlike some of his contemporaries such as the British biologist J. D. Bernal. But from the considerable body of his comments, observations, letters and interventions in such matters that is available some enduring themes are clearly visible. These themes indeed are not mutually contradictory and one can certainly discern a certain coherence in Einstein's social and political views.

The hallmark of Einstein's political vision was his deep and abiding commitment to the cause of peace. Einstein's contribution to the promotion of nuclear disarmament in the post-Second World War period is perhaps more generally known. But without examining the record of Einstein's pacifism in the context of the First World War, one would miss the radical mould in which his pacifism was cast.

If the Second World War had a clear moral justification, there was little such moral underpinning for the one preceding it. The Great War, as it was known in its time, was accompanied by an extreme outpouring of nationalist chauvinism in all the countries participating in the war. The high human cost of the war, marked by long drawn out positional warfare that resulted in huge casualties without any substantial military gain and the introduction of chemical warfare in the form of poison gas, bred in turn a growing radical opposition to the conflict. Opposition to the war and the espousal of pacifism implied a political position that, in part at least, was associated with the radical Left in the politics of that era. The most radical opposition was in Russia, where withdrawal from the war and the signing of peace with Germany was one of the slogans of the radical movement for the overthrow of the Tsar.

Einstein's pacifism first found public expression in this context [2], where those opposing the war, on both sides of the conflict, risked being labeled traitors and attracted the punitive attention of the state. Despite the risk, in October 1914, Einstein joined a small group of academics in the University of Berlin in signing a manifesto calling for European unity. The manifesto itself was a counter to another, issued by an array of German intellectuals, including many of Germany's leading scientists (and Einstein's colleagues and friends) that defended Germany's conduct of the war in the face of allegations of atrocities by the Allies.

In November 1914, Einstein joined the New Fatherland League, an organization to promote peace and European unity, as a founding member and began to participate in its activities. The organization was subsequently banned by the German government in early 1916. It, of course, attracted the attention of the police and Einstein's name appeared on the list of pacifists that they were to keep a watch on. Remarkably, as Einstein's pre-eminent scientific biographer, Abraham Pais, notes, this was also the period when Einstein was at the height of his scientific prowess, completing his formulation of general relativity, and publishing no less than fifty papers during the war years.

Einstein's political vision, as expressed in this early political activism, continued to be further sharpened in the period between the two World Wars. By the early 1930s he had moved to a critical view of the nationalist state, identifying the maintenance of national armies as playing a key role in promoting militarism. In a Gandhian vein, he emphasized the importance of individual moral commitment as a political act in the resistance to war, urging individual refusal to participate in military service, including compulsory military service in peace-time [3]. Einstein's perception of nationalism as providing the ideological justification for militarism and thus encouraging the preparedness for war, was to find later expression in his calls for a world government to halt the spread of nuclear weapons.

Nothing expresses more forcefully the radical, indeed revolutionary, cast of Einstein's pacifism than the following words from a 1931 article: "There are two ways of resisting war — the legal way and the revolutionary way. The legal way involves the offer of alternative service, not as a privilege for a few, but as a right for all. The *revolutionary view involves uncompromising resistance, with a view to breaking the power of militarism in time of peace or the resources of the state in time of war* ... both tendencies are valuable ... certain circumstances justify the one and certain circumstances the other."(emphasis added) [4].

Einstein certainly did not suffer from the weakness of converting his own beliefs and opinions into dogma. In the face of Nazism, he recognized the need to resist it by force of arms if necessary. With the accession of Hitler to power in Germany, he recognized the need of other European nations to arm themselves to resist fascism. If in his earlier emphasis on the importance of individual commitment to refusing military service he had echoed Gandhian views, he made a sharp departure from it now with his view that conscientious objection was an inappropriate policy in the face of the fascist threat.

Einstein returned to his pacifism after the war. Deeply unhappy at the bombing of Hiroshima and Nagasaki (he was to say later that Roosevelt would not have permitted it if he had been alive [5]), Einstein returned more insistently than ever before to the theme of ensuring peace in the future by handing over the control of nuclear weapons to a supranational government or organization. Einstein initially hopefully urged the government of the United States in this direction, appealing for co-operation with the Soviet Union, and was even willing to countenance the manufacture of nuclear weapons by the U.S. in the interim period. But he was soon disappointed and turned sharply critical of United States policy. Despite his criticism of the Soviet opposition to his proposals, he nevertheless acknowledged the failure of the United States to deliver any credible assurance to the Soviet Union regarding its security that would have encouraged the latter to co-operate in the search for some means of supranational control of nuclear weapons. He was particularly critical of the unwillingness of the United States to guarantee the 'no first use' (as it would be known in current nuclearspeak) of nuclear weapons.

By 1952, Einstein had returned once again to an absolute pacifism, more absolute perhaps than his position in his inter-war years [6]. He argued for the radical abolition of all wars and the threat of wars by agreement between nations as the only solution to the problem of peace rather than trying to limit the means by which wars were waged. "One has to be resolved," he wrote,"not to let himself be forced into actions that run counter to this goal." As for the means of achieving this goal, he turned once again to the example of Gandhi's leadership of the Indian freedom struggle, citing it as an example of "how a will governed by firm conviction is stronger than a seemingly invincible material power."

We may note here that Einstein certainly considered that scientists and technologists had a particular moral responsibility, especially in the era of nuclear weapons. Einstein himself was at the forefront in mobilizing his scientific colleagues on questions of peace and disarmament. But it is also evident that his own pacifism was rooted in a political and moral standpoint that went much deeper than the question of the social responsibility of scientists and technologists alone.

In the light of Einstein's radical pacifism it is unsurprising that Einstein, as the years went by, was an increasingly powerful voice in defending the rights of individuals and groups against the power of the state. Einstein was not in the least hesitant to use his considerable prestige and influence to speak up on behalf of those whom he saw as standing up to the tyranny of the state. One of the first such interventions by Einstein was his appeal on behalf of Friedrich Adler, a notable radical socialist leader in his day and a fellow-student and friend from Zurich. In 1916 Adler assassinated the Minister-President of Austria, notorious for his authoritarian rule. Einstein readily offered to intervene on his behalf [7] and publicly defended his friend [8].

Einstein's readiness to defend the right to the freedom of thought and expression was to find full expression in his defence of those who were the targets of the anti-Communist witch-hunting in the United States of the McCarthy era. Einstein's defence of the American Left in a period of sustained attack on their rights is one of the many examples from this period. Einstein did not hesitate to be publicly associated with known American communists such as the singer Paul Robeson and the historian and civil rights leader W. E. B. Du Bois [9]. Einstein turned again to Gandhi and advocated "revolutionary non-cooperation in the sense of Gandhi's" [10] as the only option for the intellectuals who were sought to be intimidated by being hauled up before U.S. Congressional committees. If intellectuals were not prepared to resist this intimidation, "then the intellectuals of this country deserve nothing better than the slavery that is intended for them."

Einstein also devoted considerable attention to the question of racism [9, 11]. In 1946, a year marked by racist incidents including several lynchings involving returning black American soldiers, he despatched a letter to President Truman, asking for the passage of an anti-lynching law. Einstein's home in the United States, Princeton, was itself steeped in racism. As late as 1942, Princeton University refused to admit black students as white Southern students would find it offensive. In this atmosphere, Einstein made his own anti-discrimination stance amply clear. When in 1937, a noted black opera singer could not find accomodation in Princeton town, Einstein invited her home to stay with him, thus beginning a friendship that was to continue to the end of Einstein's life.

As he had demonstrated continually from the days of the First World War, Einstein never lacked in personal courage when it came to speaking up for freedom. If on the one hand his own enormous prestige gave him ample protection, Einstein on the other hand went farther than most of his eminent contemporaries in speaking up against any form of authoritarianism. Einstein's prestige was unable to protect him only when it came to Nazism, leading to Einstein's early departure from Germany before the Nazis seized power. In the United States, Einstein was often the subject of right-wing attacks in the United States and the target of editorial criticism even in newspapers such as the New York Times and the Washington Post in the McCarthy era. As is publicly known today, the Federal Bureau of Investigation of the United States Government, under its notorious head, J. Edgar Hoover, continually spied on him throughout his entire life in the United States [12]. It is unclear whether Einstein was aware of the FBI's surveillance, but it undoubtedly had little effect on him or his political activism.

Einstein's courage in defending the right to the freedom of expression is all the more remarkable for the great lack of it that characterized academic life, particularly in the sciences, in the United States even in the post-McCarthy era. In Einstein's own discipline of physics, the leading figures of the next generation were noteworthy for their political conformism and readiness to collaborate with U.S. militarism. Even after the upheavals of 1968, it was not until the disastrous end of the Vietnam war that attitudes began to change. Only in the era of the Star Wars programme, after the great anti-nuclear protests of the 1970s, did opposition to nuclear militarism became respectable again in mainstream physics circles in the United States.

Einstein's radical pacifism extended also to a close interest in socialist political thought. The Zurich of Einstein's student days was home to an array of socialist leaders and thinkers, many of them exiles from their homeland. Socialism was very much in the air as an ideology and Einstein must not have been unaware of the different currents of political thought swirling around him. His friendship with Friedrich Adler, referred to earlier, would have certainly occassioned some exposure to socialist thought, since Friedrich, apart from being an active socialist himself, was the son of one of the leading Austrian social-democrats of his day, Victor Adler [13].

Einstein was certainly sympathetic to the socialist experiment in the Soviet Union. For almost a decade he was part of the central committee of a German organization to promote public knowledge about developments in the Soviet Union, the "Society of German friends of the new Russia" [14]. Though Einstein appears to have never taken the Soviet state to task publicly in strong terms on the question of freedom of speech and expression, it is likely that he would have been critical of the Soviet state on that score. But nevertheless he was clearly attracted to the economic aspects of socialist thought. The most detailed account of Einstein's views on socialism comes from the little essay titled 'Why Socialism' that he wrote for the inaugural number of the American Marxist journal Monthly Review in 1949 [15, 16].

Einstein's presence as an eminent contributor in that inaugural issue is itself an interesting illustration of the close contact that Einstein maintained with left-wing political circles [9]. The journal was founded by Leo Huberman and Paul Sweezy, who had both participated in the U.S. presidential campaign of the Progressive Party candidate Henry Wallace in 1948. The party, formed from the left-wing of Roosevelt's New Deal coalition, had a number of those who felt that a clearer socialist position should have been articulated in the campaign and the founders of the Monthly Review were among these. Einstein personally endorsed and supported the Wallace campaign. He was invited to contribute to the inaugural number of the new journal by Huberman's friend Otto Nathan, who was himself a left-wing social democrat and a close friend of Einstein (Otto Nathan and Einstein's secretary Helen Dukas were the two trustees named in Einstein's will).

It is clear from Einstein's essay that he is attracted by the socialist critique of capitalism. In characteristic fashion though, Einstein's attraction to this critique is founded in his preoccupation with the role of the individual in society. For Einstein, the individual is, as he argues at length in the first part of the essay, very much a social being. While this fact is unalterable, society itself and thus the 'cultural constitution' that individuals acquire from society (as distinct from the 'biological constitution' that human beings acquire from nature as a species) is susceptible to change.

It is this that gives hope that human life may be made more tolerable by striving to change it in a desirable fashion.

In Einstein's view the "essence of the crisis of our time" is rooted in the individual's relationship to society. Though individuals are ever more dependent on society as a whole and ever more conscious of this dependence, they are unable to perceive this as a 'positive asset' or a 'protective force'. Instead this relationship is perceived as a threat to one's natural rights or even economic existence, leading to a progressive deterioration of an individual's social drives. The real source of this alienation, this "crippling of the social consciousness of individuals" as Einstein puts it, is in his view due to the economic anarchy of capitalism and the attitudes that it promotes. Only the elimination of the anarchy of capitalism by the establishment of a socialist economy, together with an educational system that is oriented towards social goals rather than inculcating an "exaggerated competitive attitude" could solve the crisis of the individual in society.

Einstein is careful to note that socialism does not automatically follow once the anarchy of capitalism is eliminated by the establishment of a planned economic system that produces for use and not for profit. For Einstein, as he indicates in the closing lines of the essay, the problem of countering the power of the bureaucracy associated with a planned economy and safeguarding the rights of the individual remains an unsolved one. This is clearly a reference to the Soviet Union of his time and his view that socialism, in the sense of doing away with the alienation of the individual in society, was yet to be established there.

The other notable political issue that occupied Einstein was the Palestine question [17]. Einstein had considerable sympathy for the idea of a Jewish homeland even before the Second World War and was an active supporter of the Zionism of the pre-War period. Though he was born into an irreligious Jewish family, the adult Einstein clearly felt his Jewishness keenly. Einstein always maintained a distanced and objective view of other denominational religions. But he tended to a softer approach on Judaism, viewing it on occasion as rather more of a cultural expression of a particular community than a religion. In this period he was also concerned with promoting Jewish-Arab unity, spoke out against Jewish extremism and was not in favor of the partition of Palestine.

Subsequent to the horror of the Holocaust (among the victims were several of his relatives and two of his cousins), he was an even more active supporter of the idea of a Jewish homeland in Palestine. But as late as 1946–1947, Einstein continued to argue for a Palestine that would be governed under the trusteeship of the United Nations with a constitution that guaranteed that neither the Jews nor the Arabs would outvote each other and disclaimed any sympathy for the idea of an exclusive Jewish state. Paradoxically though, at the same time he maintained his close relations with the Zionist leadership that certainly had a different agenda.

Subsequently when the United Nations ended the British mandate in Palestine and violence broke out between Jews and Arabs, Einstein appealed for an end to fanaticism and violence. With the creation of the state of Israel, Einstein accepted it as a *fait accompli*, and worked to assist the new state, particularly in its scientific development. Mercifully, Einstein passed away before the Palestine question acquired the complexity that we see today.

The record of Einstein's involvement in public affairs and his engagement with the foremost political and social questions of his age is one of remarkable consistency and courage. As in his science, he was not to choose the comfort of conformism. In the pursuit of his political commitments, Einstein was willing and able to engage with the world at large in a manner that had few parallels amongst his colleagues. Einstein, as numerous personal accounts testify, was as at ease in the company of radical political activists who devoted their lives to the cause of the underprivileged, as he was in the company of statesmen and world leaders.

From the question of world peace to the crisis of the individual's relation to society under capitalism, many of the political and social issues that Einstein sought to address continue to be important questions even today. The manner in which he attempted to engage with these issues is no less inspiring today than it was to the world in his day and age.

References

[1] Schlipp, P. A. (ed.) [1949] *Albert Einstein: Philosopher-Scientist*, The Library of Living Philosophers Inc., Evanston, Illinois, USA.

[2] The record of Einstein's pacifism in the First World War that is presented here is drawn from Pais, A. [1994] *Einstein Lived Here* (Clarendon Press, Oxford) 165–169.

[3] Pais, A., op.cit., 174–176.

[4] Pais, A., op.cit., 178.

[5] Pais, A., op.cit., 233.

[6] Einstein, A. [1979] On the abolition of the threat of war, in *Ideas and Opinions* (*Indian Ed.*) (Rupa and Co., New Delhi) 165–166.

[7] Hey, T., Hey, A. J. G. and Walters, P. [1997] *Einstein's Mirror* (Cambridge Univ. Press) 6.

[8] Abstract of lecture by Galison, P. at the Stanford University web page http:www.stanford.edudeptnewspr2004pr-hofstadter-051805.html.
[9] Simon. J. [2005] Albert Einstein, Radical: A political profile, *Monthly Rev.*, May.
[10] Pais, A., op.cit., 238.
[11] Jerome, F., The hidden half-life of Albert Einstein: Anti-racism, *Socialism and Democracy Online*, Vol. 33, available at http:www.sdonline.org33/fred_jerome.htm.
[12] Jerome, F. [2002] The Einstein file: J. Edgar Hoover's secret war against the world's most famous scientist (Saint Martin's PressGriffin, New York).
[13] Datta, K. [2005] The early life of Albert Einstein: Seeking the Mature Einstein in his youth, *Resonance*, 85.
[14] Pais, A., op.cit.
[15] Available at the website of the *Monthly Review* at http://www.monthly-review.org/598einst.htm
[16] Einstein, A., Why socialism? in *Ideas and Opinions*, op.cit., 151.
[17] The material on Einstein's views on the Palestine question is based on the material in *Ideas and Opinions*, op.cit., 171–204, and Pais, A., op.cit., 242–252.

Einstein, Yukawa, Wheeler and Bhabha at Institute for Advanced Study, Princeton (Courtesy: Tata Institute of Fundamental Research, Mumbai)